The ABCs of Global Warming

The ABCs of Global Warming

What Everyone Should Know About the Science, the Dangers, and the Solutions

by
Charles Siegel

Omo Press

adolescentium alunt
senectutem oblectant

Published by:
Omo Press
Berkeley, California
omo@omopress.com
www.omopress.com

Publisher's Cataloging-in-Publication Data

Names: Siegel, Charles, author.
Title: The ABCs of global warming : what everyone should know about the science , the dangers , and the solutions / by Charles Siegel.
Description: Includes bibliographical references. | Berkeley, CA: Omo Press, 2021.
Identifiers: LCCN: 2020904271 | ISBN 978-1-941667-19-4
Subjects: LCSH Global warming. | Climatic changes. | Global temperature changes. | Greenhouse effect, Atmospheric. | BISAC SCIENCE / Global Warming & Climate Change
Classification: LCC QC981.8.C5 .S54 2020 | DDC 333.7/14--dc23

Contents

Part 1
The Science Is Clear

It is easy to understand the basics of the science of climate change. We will begin with a quick overview. Then we will go on to look at more details.

Overview of the Science

A greenhouse heats up because its glass lets in the sun's rays but blocks some heat from escaping. Likewise, scientists have known since the nineteenth century that carbon dioxide and other greenhouse gases let the sun's rays into the atmosphere but block some of the heat from escaping.

Human activities have released a huge amount of carbon dioxide in the atmosphere. The two largest sources are:

- **Fossil Fuels:** Carbon is stored in coal, gasoline, and natural gas. These are called fossil fuels because they were formed from organisms that died millions of years ago, which removed carbon dioxide from the atmosphere as they grew. When we burn fossil fuels, their carbon is released into the atmosphere again as carbon dioxide.

- **Deforestation:** Carbon is stored in trees, which removed carbon dioxide from the air as they grew. When we clear forests for agriculture, the trees rot or burn, and the carbon is released into the atmosphere again as carbon dioxide.

For 800,000 years, the Earth's climate has gone through a natural cycle alternating ice ages and warmer interglacial periods. During ice ages, the concentration of carbon dioxide in the atmosphere went down as low as 180 parts per million (ppm); during the warmer periods the concentration of carbon dioxide went up as high as 290 ppm. Notice that the difference between the coldest and warmest points in this cycle is 110 ppm. The percentage increase determines the amount of warming, and there was an increase of a bit over 60%.

At the beginning of the industrial revolution, in 1750, the concentration of carbon dioxide was 280 ppm; now it is about 410 ppm—an increase of about 45%—and it is still rising. Emissions have been increasing so quickly that we have added about as much carbon dioxide in the last 40 years as we added in the first 230 years of the industrial revolution. And human activity has also added other greenhouse gases besides carbon dioxide.

As expected, the increase in carbon dioxide and other greenhouse gases has increased global temperatures. Surface temperatures have risen by just over 1° Celsius (1.8° Fahrenheit) and ocean temperatures by just under 1° Celsius since the beginning of the industrial revolution.

Figure 1 shows that temperatures increased as carbon dioxide concentrations in the atmosphere increased. From 1880 to the 1930s, the average annual temperature was always below the average for the entire period from 1880 to the 2000s. From the 1930s through the 1970s, annual temperature was sometimes below and sometimes above the average. Since 1980, it has always been above the average. The long-term trend is very clear, though there are fluctuations from year to year because of events such as volcanoes (which lower temperatures by reflecting sunlight into space), La Niña (which lowers surface temperature as cool water wells up from the ocean depths off the coast of South America), and El Niño (which raises temperature as warm water wells off he coast of South America).

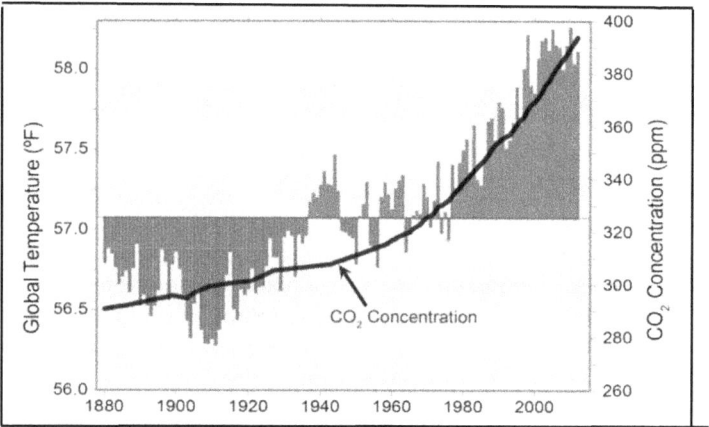

Figure 1: Carbon Dioxide and Global Temperature[1]

Global warming has already caused more destructive tropical storms, worse flooding, and worse wildfires. These effects are noticeable now, with a temperature increase of about 1°C. If we do not limit global warming, there will be much worse damage in the future.

In the Paris agreement of 2015, the nations of the world reached consensus that we should limit warming to well below 2°C in order to avoid the most destructive effects of global warming.

This overview should make the basic facts about global warming clear. The key facts are:

- Scientists have known since the nineteenth century that carbon dioxide traps some of the sun's heat.
- Human activity has released vast amounts of carbon dioxide and other greenhouse gases.
- Since the beginning of the industrial revolution, the concentration of carbon dioxide in the atmosphere has increased from 280 ppm to 410 ppm.
- At the same time, the Earth has become warmer.

Keeping these simple facts in mind can immunize us against attempts to deny or obfuscate the science of global warming. Now, let's look at the science in more detail.

Setting The Earth's Temperature

The Earth is heated by the sun, but most of that heat goes back into space. When we increase the concentration of greenhouse gases in the atmosphere, we trap more of that heat, so the Earth becomes warmer.

The Greenhouse Effect

About 29% of the solar energy that reaches the Earth is reflected back into space by clouds and aerosols in the air or by ice, snow or other bright areas of the Earth's surface. (This reflectiveness is called the "albedo effect.")

The rest of this solar energy is absorbed by the air or the Earth's surface—about 23% by the air and about 48% by the surface.

Most of the energy that is absorbed by the surface leaves through evaporation (about 25%), convection (about 5%), and radiation of infrared energy (about 17%).

To understand the greenhouse effect, we have to look at the 17% of solar energy that leaves the surface as infrared radiation.

A tiny portion of the atmosphere is made up of gases that absorb some of the infrared radiation escaping from the surface. Less than 1% of the atmosphere is made up of carbon dioxide and other greenhouse gases that trap some of the energy of the infrared radiation. Over 99% of the atmosphere is made up of nitrogen, oxygen, and argon, which do not trap infrared radiation.

Of the 17% of total solar energy that leaves the surface through radiation, about 12% escapes into space, and about 5% is absorbed by these greenhouse gases, warming the atmosphere.

This is called the "greenhouse effect," given this name because the greenhouse gases let sunlight enter but prevent some heat from escaping, a bit like a greenhouse.

The greenhouse effect increases the Earth's average temperature to about 15°C (59°F). Without it, the average temperature would be about 30°C lower—about -15 °C (5°F), well below the freezing point—and the Earth would be much less capable of supporting life than it is now.[2]

Global Warming

The amount of energy that the Earth's surface radiates increases as temperature increases.

If more greenhouse gases are added to the atmosphere, more heat is trapped, so the Earth's temperature increases, which also means that the amount of heat the Earth's surface radiates increases. The temperature rises until it reaches the point where the increased amount of heat radiated into space is equal to the increased amount of heat trapped by the added greenhouse gases, and then the temperature is stable again.

But if we keep adding more and more greenhouse gases, the temperature will keep going up higher and higher—and we have been adding more and more. Between 1750 to 2011:

- **Carbon dioxide** concentration in the atmosphere increased by 40%.

- **Methane** concentration increased by 150%.

- **Nitrous Oxide** concentration increased by 20%.[3]

Soon, we will look at all the greenhouse gases generated by human activity in more detail.

Climate Before the Change

To put our current situation in perspective, it is useful to look at the climate before human activity began to change it.

Ice Ages and Warmer Periods

For the last 800,000 years, the Earth's climate has gone through a natural cycle alternating glacial periods, when the concentration of carbon dioxide in the atmosphere went down as low as about 180 ppm and warmer interglacial periods when the concentration of carbon dioxide went up as high as about 290 ppm. The glacial periods are commonly called ice ages.

The Earth has been in an interglacial period called the Holocene for about 11,700 years.

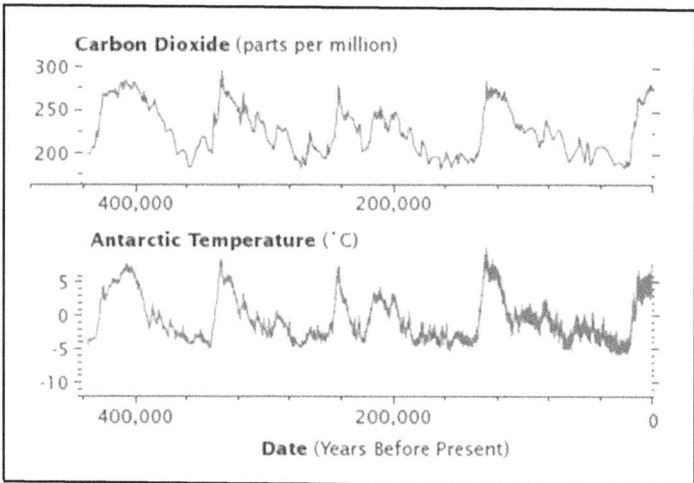

Figure 2: Carbon Dioxide During Ice Ages and Warmer Periods[4]

This cycle is caused by irregularities in the Earth's orbit and axis that occur periodically. Ice ages begin when these changes cause a bit of cooling, which leads to a self-reinforcing process that makes it get colder and colder:

• The slight cooling makes more polar ice form and makes the oceans absorb more carbon dioxide,

- which makes it even cooler by increasing the albedo effect of the ice and reducing the greenhouse effect of the carbon dioxide,
- which makes even more ice form and makes the oceans absorb even more carbon dioxide,
- which makes it even cooler,
- and so on as the process keeps reinforcing itself.

During the last ice age, carbon dioxide levels went down to about 180 ppm, and glaciers extended so far south that they covered the entire northern edge of the United States, carving out features such as the Finger Lakes of New York State.

This self-reinforcing process is reversed when the Earth's orbit shifts and the axis wobbles in ways that make it a bit warmer. Then some ice melts and oceans release some carbon dioxide from the into the atmosphere, making it even warmer, causing more ice to melt and releasing more carbon dioxide, and so on. The self-reinforcing process continues, the glaciers retreat, and the climate of the temperate zone becomes temperate again

It is useful to look at this deep history to see how relatively small changes in temperature can lead to much larger ones, partly because of changes in amount of ice on the Earth and partly because of changes in carbon dioxide concentrations. Figure 2 shows how closely changes in temperature are correlated with changing levels of carbon dioxide.

It is also useful to see that small differences in carbon dioxide levels make a big difference in climate. The difference between the ice ages and the warmer periods is 110 ppm—just over .01% of the atmosphere.

The Old Normal

Our species, homo sapiens, evolved under these conditions. All of the crops and animals we rely on for food were domesticated during the last 10,000 years, after the end of the last ice age.

Humans and our sources of food are adapted to this stable climate, but this old normal is now changing into a new normal, warmer than the climate that we and our food sources are adapted to.

Human-Caused Emissions

Carbon dioxide is the most important greenhouse gas that we are adding to the atmosphere, but there are also others.

Carbon Dioxide

Since the beginning of the industrial revolution, human activity has added vast amounts of carbon dioxide to the atmosphere, primarily as a result of burning fossil fuels. The graph shows the largest sources of carbon dioxide emissions, which we can see are approaching 40 billion tons per year and are increasing rapidly.

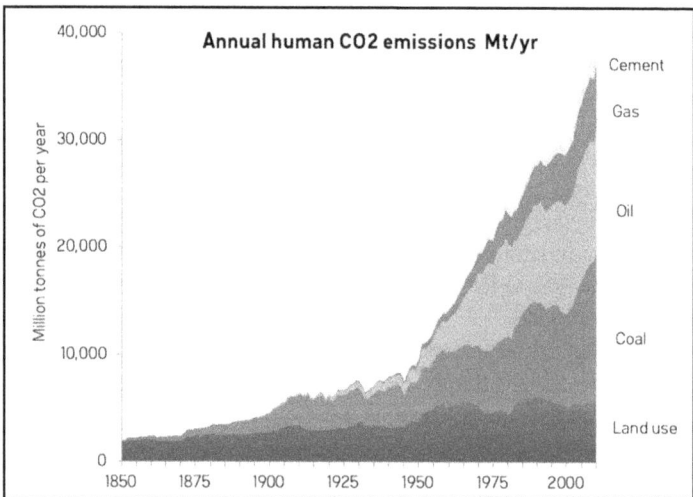

Figure 3: Carbon Dioxide Emissions[5]

Of the total carbon dioxide emissions:

- **About 40%** remains in the atmosphere, causing warming.
- **About 30%** is absorbed by the oceans, causing them to be more acidic, as we will see.
- **About 30%** is absorbed by vegetation and soils.[6]

Greenhouse Gases

Carbon dioxide is not the only problem. Here is an overview of all the greenhouse gases:

- **Carbon dioxide (CO_2)** accounts for about 76% of current emissions.[7] As we have seen, it makes up about 410 parts per million of the atmosphere, which is an increase of about 45% since the beginning of the industrial revolution.

- **Methane (CH_4)** is produced when organic matter decays in the absence of oxygen, for example, in landfills, in cows' stomachs, and underground (where large amounts are stored as natural gas). It makes up about 1.7 ppm of the atmosphere, 150% more than it did before the industrial revolution. It lasts 12.4 years before combining with oxygen in the air to form carbon dioxide and water vapor. Over 20 years, it creates 67 times as much warming as the same volume of carbon dioxide.

- **Nitrous oxide (N_2O)** is emitted naturally by microorganisms in soils and the oceans and also results from human activity, primarily from nitrogen fertilizers, animal and human manure, and burned biomass. It is used in medicine and is popularly called "laughing gas." It makes up about 0.330 ppm of the atmosphere, 20% more than it did before the industrial revolution. It lasts 121 years. Over 20 years, it creates 277 times as much warming as the same volume of carbon dioxide.

- **Halocarbons (CFCs and HFCs)** do not exist in nature; they are artificial gases used in air conditioning and refrigeration. Hydrofluorocarbons (HFCs) replaced chlorofluorocarbons (CFCs), which were phased out because they damage the ozone layer. But HFCs do contribute to global warming. Some can create thousands of times as much warming as the same volume of carbon dioxide. Concentration in the atmosphere ranges from 1 part per trillion (ppt) to 65 ppt. HFC152a is now commonly used as a refrigerant because it causes less global warming than most HFCs. It lasts 1.5 years, and over 20 years creates 174 times as much warming as the same volume of carbon dioxide.

- **Tetrafluoromethane (CF4)** is used as a low-temperature refrigerant for manufacturing electronic components. It lasts 50,000 years. Over 20 years, it creates 5270 times as much warming as the same volume of carbon dioxide. Because of the tiny amount in the atmosphere, it contributes little to global warming.[8]

- **Water Vapor (H_2O)** traps heat and accounts for about 0.5% of the atmosphere, much more than carbon dioxide. But unlike other greenhouse gases, which remain in the atmosphere for a long time, water vapor falls to earth as rain or snow, so it does not accumulate in the atmosphere like other greenhouse gases. However, the atmosphere can hold more water vapor when it becomes warmer, and this added water vapor increases the amount of warming. For example, if we added enough carbon dioxide to the atmosphere to cause 1°C of warming in itself, we would actually get about 3°C of warming, because the warmer atmosphere would hold enough added water vapor to cause about 1.5°C of additional warming, and enough ice would melt to cause about 0.5°C of additional warming. Thus, the other greenhouse gases cause warming, and water vapor amplifies the warming.

Because carbon dioxide is the most important greenhouse gas, we often describe the effect of these other gases based on "carbon dioxide equivalent," the amount of carbon dioxide that would cause the same amount of warming. Currently the effect of all the other gases we have emitted is equivalent to the effect of about 75 ppm of carbon dioxide.[9]

Aerosols

Apart from these gases, human activity generates two aerosols (tiny particles suspended in the air) that have an important effect on global warming.

- **Sulfate aerosols** come from volcanoes, biological sources, and from burning forests and fossil fuels. They reflect the sunlight that hits them, cooling the Earth, which is why major volcanoes cause temporary cooling. They remain in the atmosphere for 3 to 5 days. It is estimated that there are now more human caused sulfate aerosols, mostly from burning coal and oil, than naturally caused aerosols, which means that human activity has increased them by over 100%. They are concentrated in the northern hemisphere, where most industrial activity is. They cool the Earth about as much as reducing carbon dioxide levels by 60 ppm.[10]

- **Black soot or black carbon** is emitted by burning coal or diesel fuel and by burning wood or dung for cooking and heating in the developing nations. Because it is black, it absorbs sunlight, warming the atmosphere. It remains in the atmosphere for days or weeks. The Intergovernmental Panel on Climate Change (IPCC), which is the most reliable source of information about global warming, says there are large uncertainties about its effect on global warming.[11] One recent study found that it absorbs twice as much heat as earlier estimates, which would make it the second greatest cause of global warming, totaling about two-thirds as much impact as all carbon dioxide emissions.[12]

Self-Reinforcing Warming

The warming that has already occurred has started natural processes that are self-reinforcing and that could lead to runaway global warming if they are not stopped. For example:

- **Arctic ice** now covers only half as much area after the summer melt as it did in 1979.[13] The white ice reflects the sunshine, and when it melts it exposes the darker ocean surface that absorbs the sun's energy—which causes more warming, which causes more ice to melt, and so on.

- **Forests** are expected to die back in many regions because of global warming,[14] as a warmer climate lets insects move further north and makes some regions too dry to support forests. Global warming also causes worse forest fires. As trees decay or burn, they release more carbon dioxide—which causes more warming, which causes more trees to die, and so on.

- **Peatlands** are wetlands that are waterlogged year round, slowing the decomposition of organic matter, which forms peat. They store 550 billion tons of carbon, more than the world's forests. Global warming will dry some peatlands, releasing the carbon stored there either as methane or as carbon dioxide.[15]

- **Permafrost** contains large amounts of methane. Warming causes permafrost to melt,[16] releasing methane—which causes more warming, which causes more permafrost to melt, and so on. As permafrost melts, organic matter that was trapped in it begins to decompose, generating heat that accelerates the melting. This process is very dangerous, because there are about 1,400 billion tons of carbon trapped in the permafrost, far more than the 850 billion tons in the atmosphere,[17] and as the permafrost melts, this carbon is released as methane, which causes 67 times as much warming as the same volume of carbon dioxide over the first 20 years.

Rising Temperatures

Increased concentrations of carbon dioxide, other greenhouse gases, and black soot have already increased temperatures by about 1°C.

The data makes it very clear that temperature is increasing as greenhouse gas concentrations are increasing. The graph in the overview section shows that, though there are fluctuations from year to year, the long-term trend is for temperatures to increase as carbon dioxide concentrations increase.

The Science and the Data

The basic science is very clear. Scientists have known since the nineteenth century that carbon dioxide traps heat in the atmosphere.

In the 1850s, John Tyndall did experiments to measure how much radiant energy the gases in the atmosphere can absorb. He filled tubes with each of the gases in the atmosphere and subjected the tubes to radiant heat, and he found that carbon dioxide and water vapor trap heat, while oxygen and nitrogen do not.[18] In 1896, the Swedish scientist Svante Arhenius used the principles of physical chemistry to estimate how much carbon dioxide increases the Earth's temperature. At first, he was trying to explain the difference in temperatures between ice ages and interglacial periods, but his calculations led him to believe that human carbon dioxide emissions could change the Earth's climate.

Now we have the data to prove this scientific theory. Carbon dioxide has increased from 280 ppm before the industrial revolution to over 410 ppm today, and temperatures have gone up. The 1980s set a record for the warmest decade in history. Then the 1990s set a new record. Then the 2000s set another new record. Then the 2010s set another new record. Four decades in a row set new records for the hottest decade in history, each one breaking the record set by the previous decade.[19]

There is a range of estimates of how much of current warming was caused by humans, but the best estimate is that, from 1951 to 2010, human activities caused all observed increases in surface temperatures.[20]

If anyone questions the basic science of global warming, ask if they can deny these facts:

- Carbon dioxide has been shown experimentally to absorb infrared radiation and become warmer.
- Human have increased the amount of carbon dioxide in the atmosphere by burning fossil fuels.
- The Earth's average temperature has increased.

Stabilizing Temperatures

How much temperature increases in total depends on how much greenhouse gas is emitted in total, which depends on how quickly emissions decline and are reduced to net zero. The IPCC recently found that:

- To limit warming to 1.5°C, human-caused carbon dioxide emissions must reach net zero around 2050.
- To limit warming to 2°C, human-caused carbon dioxide emissions must reach net zero around 2070.

In both cases, there must also be large reductions in emissions of methane and black carbon, though they do not have to reach zero by 2100.[21]

To understand why it is essential to limit warming, we need to look at the dangers of global warming and at how much worse these dangers become when warming exceeds 1.5° to 2°C.

Part 2
The Dangers Are Frightening

We can already see effects of global warming, such as worse heat waves, more destructive tropical storms, more flooding, and larger wildfires. Things will get even worse as warming continues.

Heat Waves

It seems obvious that rising global temperatures will make extreme heat waves more common, and it has already happened. With the 1°C temperature increase that we have so far, there are already monthly heat records five times more frequently than they would without warming.[22] Here are a few examples:

- In 2003, Europe had the worst heat wave ever recorded, which caused 35,000 people to die, breaking the record for deaths caused by heat.[23]

- In 2010, Moscow had one of the worst heat waves on record, which worsened forest fires. The combination of heat and smoke caused 11,000 deaths.[24]

- In 2019, India and Pakistan had a heat wave that sent temperatures soaring to 123° F (50.5°C) and killed 100 people.[25] This heat wave came in the wake of predictions that, if the world does not control global warming, parts of south Asia, including the Indus and Ganges valleys where hundreds of millions of people live, will be too hot for human habitation by the end of this century.[26]

- In 50 major American cities, the number of heat waves has tripled since the 1960s.[27]

Apart from causing deaths, extreme heat waves can have severe economic effects, particularly on agriculture and construction, which require people to work outdoors. One study found that hot days decrease daily productivity in the United States by a surprising 28% compared with average days.[28]

Under the scenario where the world does not limit global warming, heat would severely impact many common human activities.[29] There would be particular risk to the health and lives of children, the elderly, and women in the developing nations who must walk long distances to gather water.[30]

Limiting warming to 1.5°C could minimize this problem. If we limit global warming to 2.0°C, there would be 420 million more people frequently exposed to extreme heat waves than if we limit warming to 1.5°C.[31]

More Destructive Tropical Storms

Tropical storms, called hurricanes, cyclones, or typhoons depending on where they occur, get their energy from the warmth of the ocean. Scientists estimate that every 1°C warming of the ocean causes a 7% increase in the maximum wind speeds of these storms. Because their destructive force is proportional to the cube of the wind speed, 1°C of warming increases the destruction they cause by 23%.[32]

The Most Destructive Storms

In recent years, with oceans that have warmed by almost 1°C, we have seen the most destructive tropical storms in history:

- **Hurricane Katrina** in 2005 and **Hurricane Harvey** in 2017 are the two most destructive Caribbean hurricanes on record, each causing $125 billion in damage. Apart from this property damage, Katrina killing an estimated 1,245–1,836 people when the levees of New Orleans failed, flooding the entire city[33] Harvey affected Texas, Louisiana, the Caribbean, and Latin America, killed at least 107 people, and caused catastrophic flooding in Houston, where there was more than 40 inches of rain in four days.[34]

- **Hurricane Patricia** in 2015 was the most destructive hurricane on record in the north Pacific and had the highest wind speed of any tropical storm on record. In Mexico, it caused $325 million in damage and six deaths; 5,791 Marines were sent to Mexico to assist in the recovery. In the United States, it caused $50 million of damage.[35]

- **Hurricane Irma** in 2017 was the most destructive hurricane on record in the open Atlantic, causing $77.2 billion in a dozen Caribbean nations and the United States. In west and central Florida alone, it damaged or destroyed more than 65,000 buildings and caused a $2.5 billion loss to agriculture.[36]

- **Cyclone Winston** in 2016 was the most destructive hurricane on record in the southern hemisphere, causing $1.4 billion in damage. In Fiji alone, it damaged or destroyed 40,000 homes and affected 350,000 people, about 40% of Fiji's population.[37]

These examples show that tropical storms have set records for destruction throughout the world—in the Caribbean, North Pacific, open Atlantic, and southern hemisphere—though the property damage is relatively small in some cases, when the storms occurred in places with little population.

In addition, other Caribbean and Atlantic hurricanes that have hit populated areas have been very damaging:

- **Hurricane Maria** in 2017 caused $91.6 billion in damage in Puerto Rico, US Virgin Islands, and Dominica;[38]
- **Hurricane Sandy** in 2012 caused about $70 billion in damage, moving up the eastern seaboard of the United States and causing $65 billion of its damage in 24 states from Florida to Maine.[39]
- **Hurricane Wilma** in 2005 caused $27.4 billion in damage, moving from the Caribbean up the Atlantic coast and causing $19 billion of its damage in the United States.[40]

And there are many more.

The Future

We are setting these records for destructiveness with 1°C of warming, which increases the destructiveness of tropical storms by 23%. It is easy to calculate that:

- If we limit global warming to 1.5°C, tropical storms will be about 34% more destructive than they were before warming.
- If we limit warming to 2°C, these storms will be about 50% more destructive than they were before warming.
- If we do not control global warming and temperatures rise by 4.5°C by the end of this century, these storms will be about 2.5 times more destructive than they were before warming—and will continue to become even more destructive as temperatures continue to rise in the next century.

Flooding and Drought

Warmer air can hold more water, which means that:

- On the average, global warming will make the land dryer overall, because more water will evaporate into the air.

- When it does rain, the rain will be heavier, since there is more water in the air to fall as rain.

Thus global warming will increase both droughts and flooding.

Effects at Different Latitudes

That is what will happen on the average throughout the world, but warming has different effects on rainfall and drought at different latitudes because of the way air circulates.

In what is called Hadley cell circulation, the hot air of the tropics rises, flows toward the poles at a height of 10 to 15 kilometers (6.2 to 9.3 miles), descends in the subtropics, and returns toward the equator near the surface. The air loses its moisture as it flows upward: at higher altitudes, the cooler temperature makes water vapor condense, creating heavy rains in the tropics. This dried air descends in the subtropics, creating a belt of deserts.[41] Global warming will cause the following changes:

- **Tropics:** With global warming, the tropical belt will expand and more water will evaporate, so there will be heavier rain as the air rises. There will be too much rain in some places: heavier rainfall in Bangladesh, for example, will worsen flooding caused by rising oceans there.

- **Subtropics:** As the tropics expand, the subtropical belts will move northward and southward. For example, the desert in North Africa will expand northward, so parts of southern Europe will become deserts. The deserts in the American southwest will move northward; Lake Powell, formed by the Glen Canyon dam, provides southern Nevada with much of its water supply but may dry up in a few decades, and Oklahoma could become a permanent dustbowl.

- **Temperate Zone:** In much of the temperate zone, there will be more precipitation—heavier rainfall and snow storms—but there will also be drier conditions most of the time because the water from the precipitation will evaporate more rapidly.

- **Poles:** Near the poles, global warming is expected to cause wetter conditions.[42]

There will also be more extremes because warmer air holds more moisture. Even the places with more rain overall are likely to have longer dry spells as well as heavier rainstorms and worse flooding.

Because global warming has these two effects—heavier rain and drier land—the portion of the world's population that will suffer from flooding and the portion that will suffer from water scarcity are both expected to increase as the world gets warmer.[43]

Heavier Rains

Global warming has already brought more extreme rainfall. The number of extreme rain events increased steadily from 1964 to 2013, at a rate of about 7% per decade.[44]

Flooding following heavy rain caused more than half a million deaths between 1980 and 2009, and also caused landslides, damaged crops, and disrupted transportation.[45] For example:

- **East Africa:** In 2018, record-breaking rains followed a year of severe drought in East Africa, displacing hundreds of thousands if people.[46] Then, in December 2019, even heavier rains destroyed crops and displaced millions of people in Somalia, Kenya, Uganda, South Sudan and Ethiopia.[47]

- **American Midwest:** The Midwest of the United States had the wettest twelve months on record from June 2018 to May 2019. Rivers overflowed and many states faced record flooding. Almost 14 million people were affected. Standing water in fields prevented planting in almost 1 million acres of farmland in nine grain-producing states.[48]

If global warming is not controlled, the number of extreme summer rains in parts of the United States could increase by 400% by the year 2100, and the average rain could be as much as 70% more intense.[49] Rain will continue to get worse in subsequent centuries if warming continues. Controlling global warming can make this impact much less severe:

- With 2°C of warming, extreme rainfall will become 36% more common than it was from 1981 to 2010.

- With 1.5°C of warming, it will become only 17% more common.[50]

Drought

In addition to worse rain, we are already seeing worse droughts:

- **Syria:** In 2006-2009, Syria had its worst drought in modern times, which devastated agriculture and herding, forcing 6 million people (more than one-quarter of Syria's population) to leave their homes. As people fled from farms to the cities, poverty and displacement helped cause the civil war that began in 2011 and that has caused over 200,000 deaths.[51]

- **Oklahoma and Texas:** In 2011, Oklahoma and Texas had their worst drought on record, which caused Oklahoma ranchers to lose 25% of their cattle.[52]

- **Southern Africa:** Since 2018 and continuing at least into 2020, there has been a severe drought in southern Africa, following flooding in 2017-2018. The famous Victoria Falls, on the border of Zimbabwe and Zambia, turned into a small stream. Vegetation died and crops failed. About 7 million people in Zimbabwe and 2.3 million people in Zambia were near starvation, and 52 million people across southern Africa were at risk of famine.[53]

If global warming continues, droughts that used to occur once every 100 years will occur once every 2 to 5 years in much of the world, including the southern and central United States, and water supply deficits in these areas will increase fivefold. Even with strict policies to limit warming, the severity of droughts is likely to double in these areas.[54]

Limiting warming to 1.5°C could minimize this problem. If we limit global warming to 2°C, there could be twice as many people suffering from water stress caused by climate change as there would be with 1.5°C of warming.[55]

Worse Wildfires

Both longer droughts and higher temperatures dry out the land, which means that wildfires spread more quickly.

In just the last few years, the extent of wildfires has increased dramatically. Worldwide, the number of people exposed to smoke from wildfires, largely in India and China, has increased 77% just since the mid 2010s.[56]

The 2016 wildfire in Alberta, Canada, burned 1,500,000 acres and was not controlled until one year and two months after it began, The cost of damage was about $9 billion, making it the most expensive disaster in the history of Canada.[57]

In California, the area burned has increased fivefold since 1972.[58] Though the number of wildfires has remained about the same, they have become much worse because they spread more rapidly in dry conditions.

The new normal in California during the last several years has been wildfires so bad that air in nearby cities is unhealthy to breathe.

2017 set a record as the most destructive year for wildfires in California, with fires burning over 1,300,000 acres of land and causing an estimated $180 billion in damage.[59] Then 2018 set a new record, with fires burning over 1,800,000 acres[60] and an estimated $400 billion in damage.[61] In 2019, only about a quarter of a million acres burned, but 2020 set another new record, with over 4 million acres burning, more than three times as much as the 2017 record and more than 4% of all the land in the state.[62]

Since 2017, there have been periods every year when smoke from these fires gave the San Francisco Bay Area some of the worst air quality in the world. The local utility, Pacific Gas and Electric declared bankruptcy in 2019 because of its liability for causing fires, and it began to protect itself from liability by shutting off electricity service when there was extreme fire danger.

One of the most technologically advanced areas in the world, the home of Silicon Valley, now seems in some ways like a backward developing nation during fire season, with unreliable electrical service and with air so bad that most people are afraid to go out without a face mask,

An Erratic Jet Stream

We have seen that many extreme events have become much more frequent because of global warming—including heat waves, flooding, storms, and wildfires. It might seem odd that the 1°C of warming that has already occurred can make these extreme events so much worse, but it can be explained in part by changes in the jet stream.[63]

The jet stream is a strong wind current that goes around the world at about 35,000 feet (10.6 kilometers) above the surface at about the latitude of the United States-Canada border, sometimes straight and sometimes meandering north and south. There is evidence that global warming is making the jet stream more irregular: it meanders further north and south.

The jet stream pushes high and low pressure areas along in its path, but it can stall when there are large meanders, so low pressure areas that can cause rain and high pressure areas that can prevent rain and cause hot spells remain in the same place for a longer time instead of being moved along by the jet stream. Because rainy or dry weather stays in the same place for a longer time, there are worse floods, hot spells, droughts and wildfires.

This change in the jet stream aggravated many recent extreme weather events. The 2003 European heat wave, the 2010 Moscow heat wave and wildfires, the 2011 Oklahoma and Texas drought, the 2016 wildfires in Alberta, all occurred at times when the jet stream stalled.[64]

In addition, the polar vortex forms during the winter in the stratosphere, and the meandering jet stream can bring it south and bring very cold arctic air with it, creating sudden dramatic changes in temperature. For example, in parts of Texas, the temperature index went from 92°F (33°C) on Monday November 11, 2019 to 31°F (-0.5°C) the next morning. The stagnant jet stream let that mass of cold air stay long enough to break records for the lowest daily temperature in history for many days on end.[65]

Melting Ice and Rising Sea Levels

Glaciers and polar ice are melting, a very obvious, visible sign of global warming. Sea levels are rising both because water expands when it becomes warmer and because melting ice is adding water to the seas. The world's average sea level has already risen .19 meters (about 7.5 inches) since 1900.[66] By the end of this century, it is expected to rise .9 meters to 1.5 or 1.8 meters (3 feet to 5 or 6 feet).[67] As sea levels rise, salt water contaminates sources of fresh drinking water, and flooding threatens coastal cities and large coastal regions.

The Arctic

Many of the early European explorers tried to find a passage north of Canada from the Atlantic to the Pacific Ocean, which they called the Northwest Passage, but they failed because the Arctic Ocean was permanently blocked by ice. The first passage was not made until 1903–1906, using a route that went south of Canada's northern islands.[68] But now, global warming has melted so much ice that hundreds of ships use routes that are further north than these islands every summer.

Since 1979, about half of the area of Arctic ice has been lost in September, the time of the maximum melt, and about 95% of the thick old ice has been lost. Almost all of the ice now is thinner new ice, so that September ice has lost about 75% of its volume since 1979. This thinner ice will melt more quickly.[69]

Because Arctic ice floats on water, it does not increase sea level when it melts, but the melting is important because speeds up global warming by reducing the albedo effect: ice that reflects solar energy into space is replaced by water that absorbs solar energy. This is why the Arctic has been warming twice as fast as the rest of the world during the last several decades.

It seems inevitable that there will be summers when all of the Arctic ice melts, and it is expected as soon as the late 2030s.[70] Limiting global warming 1.5°C could reduce their frequency dramatically:

• With 2°C of global warming, there will be ice-free Arctic summers once every ten years by 2100.

- With 1.5°C of global warming, there will be ice-free Arctic summers only once every 100 years, and the Arctic will stabilize at that level.[71]

The Antarctic and Greenland

Ice in the Antarctic and Greenland is on the land, so it adds to sea level rise when it melts.

Western Antarctica is the lowest part of the continent, so it is warmest and is melting most quickly. Parts of it have already collapsed: the Larsen B Ice Shelf on the Antarctic Peninsula, as large as the state of Rhode Island, collapsed in 2002, and the larger Larsen C Ice Shelf began to break up and collapse in 2016.

When ice shelves collapse, ice from further inland flows toward the ocean, making inland ice less stable. It is possible that, within a decade, the west Antarctic will be so unstable that most of its ice will inevitably flow to the sea and melt, causing world sea levels to rise an additional 3 meters (10 feet) beyond what is projected. This might take 1,000 years or might happen in as little as 100 years.[72]

Earlier models did not take into account this instability, so Antarctic ice is melting more quickly than scientists expected a decade ago.

For the same reason, ice in Greenland is melting more quickly than expected, with about half the loss caused by increased flow of ice. Greenland is losing ice seven times more rapidly now than in the 1990s, and is at the high end of IPCC projections.[73]

Adding ice melt caused by instability in Antarctica and Greenland to earlier predictions, seas could rise by more than 2 meters (6.5 feet) by 2100 and 6 meters (20 feet) by 2300 if we do not limit warming.[74]

The Human Costs

Cities and other densely populated areas tend to be concentrated near the coast, so vast numbers of people will be affected by rising sea levels. Many will find that their homes are below sea level, and many more will find that their homes are so close to sea level that they are flooded whenever there is heavy rain.

A recent study used more accurate calculations of cities elevations above sea level than earlier studies, correcting NASA data that used satellite measurements of the tops of buildings and trees as the level of the surface of the land. It found that, if we do not control global warming:

- By 2050, 150 million people's current homes could be below high-tide level, and an additional 300 million people would be flooded at least once a year.

- By 2100, 520 million people's current homes could be below high-tide level, and an additional 640 million people could be flooded at least once a year.

The problems would be worst in Asia, with more than 70 percent of the people affected in China, Bangladesh, India, Vietnam, Indonesia, Thailand, the Philippines and Japan.[75]

The study found that, even if global warming is limited to 2°C, the following cities and regions would be below sea level at high tide:

- Most of southern Vietnam plus Ho Chi Minh City.

- Parts of Thailand, including Bangkok.

- The center of Shanghai, China, and many areas near it.

- Much of Mumbai, one of the world's largest cities and India's financial center.

- Most of Basra, the second largest city in Iraq.

- Alexandria, Egypt, an important historical site as well as a population center. [76]

Cities that remain above sea level will be subject to more frequent flooding during storms. With just a 0.5 meter (1.6 feet) rise in sea levels, the frequency of floods would increase between ten-fold and one-hundred-fold in many locations that are near sea level,[77] including major cities like New York.

Adapting to this change will be immensely costly, as we build dykes and seawalls to save cities and other valuable land and as we move people out of less heavily populated locations. The more sea levels rise, the more likely it becomes that adaptation will become impossible[78] so cities will have to be abandoned.

We can reduce damage from sea level rise by limiting global warming, but the damage will still be vast:

- With 2°C of warming, the net present value of damage expected by 2200 is $69 trillion.

- With 1.5°C of warming, the net present value is $54 trillion.[79]

Acid Oceans

About 30% of the carbon dioxide we emit is absorbed by the oceans, where it combines with water to form carbonic acid. Some of the carbonic acid breaks down into bicarbonate ions and hydrogen ions, making the ocean more acidic.[80]

Since the beginning of the industrial era, the oceans have become 26% more acidic.[81]

Acidity reduces the availability of carbonate ions, because hydrogen ions combine with them to form bicarbonate. Organisms that use the these carbonate ions to form their shells are threatened—including commercially valuable shellfish and plankton and other organisms that are at the base of the marine food chain. Acidity also could ultimately dissolve coral reefs and shells of plankton and other organisms.[82]

The waters near California are acidifying twice as fast as the ocean generally. Shells of foraminifera (single-cell sea life like plankton) there are already 20% thinner there than they were a century ago.[83]

Along with higher temperatures, acidification is one cause of coral bleaching. Bleaching occurs because corals depend on single-celled protozoa that live in them, provide 90% of their energy, and cause their colors; when temperatures are high, these protozoa no longer provide energy and corals expel or consume them, losing their color and dying because they no longer have the nutrients that algae provide. Coral reefs are an important habitat for fish—though they occupy less than 0.1% of the world's oceans, they are the habitat of 25% of marine species—and their death could cause a major reduction in the supply of fish used as food.[84] Acidification also lowers the survival rate of fish larvae, another effect that reduces the food supply for humans.

If we control global warming and hold temperature increases to about 2°C, ocean acidity will increase by another 15% to 17% but then will begin recovering in mid-century.[85]

Extinctions

There are two well-known examples of species threatened by global warming:

- **Polar bears** go on sea ice to hunt seals, their main source of food, and now they have less area to hunt in because arctic ice is melting. Three of the nineteen polar bear subpopulations in the Arctic Circle have already declined; widespread population decline is expected by the middle of this century.[86] As we have seen, with 1.5°C of warming, the Arctic will be ice-free once every hundred years by 2100; and with 2°C of global warming, the Arctic will be ice free one year out of ten.[87]

- **Coral reefs** are becoming bleached and are dying because of warming and acidification of the ocean, as we have seen. Bleaching has already occurred in 90% of the world's largest coral reef, the Great Barrier Reef near Australia. At around 2° to 2.5°C of warming, irreversible changes will occur, as reef ecosystems disappear.[88]

These iconic species are well known, but most people don't realize how widespread the threat of extinction is. Species are moving toward the poles and moving to higher altitudes to escape warming and find climates like those they are adapted to, but many will not be able to move fast enough to keep up climate change. Species are also threatened by variations in rainfall, reduced water in rivers, ocean acidification, and lowered ocean oxygen levels.[89]

With 5°C of warming, 70% of all species could become extinct. With 4°C of warming, one-half of all species could become extinct. With 3°C of warming, one-third of all species could become extinct.[90]

Extinction can be reduced dramatically if we control global warming:

- With at 2.0°C of warming, 18% of insects, 16% of plants, and 8% of vertebrates will lose at least half their geographic range, so their populations will decline but most of them will not become extinct.

- With at 1.5°C of warming, 6% of insects, 8% of plants, and 4% of vertebrates will lose at least half their range.[91]

Poverty

The wealthier nations can spend money to adapt to climate change, but the poor people of the world will be less able to avoid its full effects. This applies to many effects we have already looked at: the poor people of the world cannot afford air conditioners to deal with heat waves and do not have insurance to compensate them if they lose their homes to fire or flooding. Here are some other ways global warming will affect the poor.

Food and Water

People dependent on local agriculture and fishing are most vulnerable:

- **Agriculture:** Local temperature increases of 2°C or more will reduce wheat, rice and maize production in tropical and temperate regions without adaptation[92]—and the poor are least able to afford that adaptation.

- **Fishing:** The oceans will become less productive. The impact will be most dramatic in the tropics, because sea life is moving toward the poles as the oceans warm; in most places, fish from warmer places arrive to replace the fish that have left, but near the equator, there are no warmer places for fish to come from, so fish stocks will collapse.[93]

Extreme weather events are also likely to have a major effect on food production, but this effect is not incorporated in the models, so the problem is probably worse than the projections say.[94]

People will also face shortages of water. In addition to droughts, there will be water shortages because of melting glaciers and reduced winter snow packs. Himalayan glaciers provide water to the Indus, Ganges, Mekong, Yangtze, and Yellow Rivers, providing water for drinking and irrigation to regions with 1.5 billion people;[95] but two-thirds of those glaciers will be gone by 2100 if we do not limit global warming, and one-third will be gone even if we limit global warming to 1.5°C.[96]

Limiting global warming 1.5°C could dramatically reduce the effect on the poor. For example:

- If we limit global warming to 1.5°C, several hundred million fewer people would be exposed to both poverty and climate risks in 2050 than there would be if we limit warming to 2.0°C.

- If we limit warming to 1.5°C, as few as half as many people would be exposed to water stress caused by climate change than there would be if we limit warming to 2.0°C.[97]

- If we limit warming to 1.5°C, there will be less of a reduction in food availability in the Sahel, southern Africa, the Mediterranean, central Europe, and the Amazon than there would be if we limit warming to 2.0°C.[98]

Climate Refugees

As global warming continues, we can expect an increasing number of climate refugees fleeing lands that are affected by drought, flooding, high temperatures, and salination of farmland caused by rising sea levels.

In 2010, in Asia and the Pacific alone, over 42 million people were displaced by storms, floods, and extreme temperatures, with more displaced by drought and rising seas. Most were able to return to their homes when conditions changed, but many were displaced permanently.[99]

The most common estimate is that the world will have 150-200 million climate refugees by 2050.[100]

How much suffering they experience depends on how the world decides to deal with them; it seems likely that nations will become less willing to welcome them as the number of refugees increases dramatically, so that many of them are likely to spend much of their lives warehoused in refugee camps.

Security

We can expect violent conflicts and civil wars as current residents compete for limited amounts of food and water and as climate refugees from other countries arrive and join the competition.

In Syria, it has already happened. That country had its worst drought in modern times from 2006 to 2009, and scientists say that such an extreme a drought was two to three times more likely because of the movement of deserts northward caused by climate change. As a result, up to 1.5 million farmers fled from the countryside to the cities, where they were economically insecure, adding to the tension that caused a rebellion against the government of Bashar al-Assad and the civil war that began in 2011, causing over 200,000 deaths and displacing over 6 million people.[101]

The same sort of thing has begun to happen elsewhere. For example, the interior of Nigeria is becoming drier, and as farms have failed, many have fled to Lagos. Likewise, as Senegal dries, many have fled and become refugees in countries that are further south, such as Ghana.[102]

This sort of displacement can cause conflicts, so climate-driven conflicts will become more common as global warming becomes worse.[103]

Uncertainty

Scientists are constantly improving their models by comparing them with actual conditions. For example, they found that ice was melting more rapidly than had been predicted a couple of decades ago because the models did not take into account the fact that, when ice melts, cracks appear and water flows down the cracks, speeding up the collapse of ice shelves; after they saw this happen, they created newer models that include the physics of how ice changes when it melts and that make more accurate predictions.

Despite all we have learned, there are still some uncertainties in the models, which mean that warming could be less destructive or more destructive than predicted.

Sensitivity and Risk

The models that the IPCC uses differ about how sensitive temperature is to greenhouse gas emissions. The IPCC's most probable estimate is that doubling carbon dioxide in the atmosphere will increase temperature by 3°C; but according to various climate models that it uses, doubling carbon dioxide could increase temperature by 1.5°C to 4.5°C. The source of this uncertainty is the effect of warming on cloud formation. It is a tricky issue: some clouds, such as high, wispy cirrus clouds, let sunlight reach the surface but block some heat from escaping, increasing warming; others, such as low, thick stratus clouds, reflect sunlight into space before it reaches the surface, reducing warming.[104]

This uncertainty is why the IPCC shows a range of projections for future temperatures, with sensitivity of 3°C as the line in the center and with a band around it wide enough to include sensitivity of 1.5°C to 4.5°C. This is also why the IPCC projections involve probabilities, saying that a given scenario of emission reductions gives us a 50% probability or a two-thirds probability of holding warming down to 1.5°C.

But what about the smaller probabilities, the results that have only 1% or 2% chance of happening?

The late Harvard economist Martin Weitzman looked at these smaller probabilities. Using sophisticated mathematical analysis based on the uncertainty of the climate's sensitivity to carbon dioxide emissions and on the possibility of melting permafrost releasing methane, he found that there is a 1% chance of 20°C (36°F) warming by 2200.

The effect clearly would be catastrophic; Weitzman estimated it could destroy 99% or more of the world's economy.[105] With this much temperature gain, the tropics would be too hot all year for people to spend time outdoors or for crops to grow. In the temperate zones, it would be too hot go outdoors during much of the year, and the summers would be too hot for crops to grow. Billions of people would die from the heat or flee toward the poles to escape the heat. The amount of food the Earth could produce would plummet. Massive displacement and hunger would undoubtedly cause war throughout the world.

So, Weitzman concluded, we should control climate change as a sort of insurance. We spend money on fire insurance for our homes, even though there is a very small chance that a given house will burn down, because we want to avoid even that small chance of being devastated financially by losing our homes. Likewise, spending money to control global warming is justified to avoid the 1% chance that warming will devastate the world economy and the Earth's ability to support human life.

Tipping points

Most impacts of global warming, such as temperature increases, get steadily worse as greenhouse-gas concentrations increase and can be reversed by reducing greenhouse gas concentrations. But there are also some impacts that become unavoidable when warming reaches some point and that cannot be reversed for thousands of years. The points where massive changes occur are called tipping points.

These tipping points involve a different sort of uncertainty. We know that are bound to reach them if warming continues, but because they are unique events, we do not have past experience to let us predict precisely when we will reach some of them. Here are examples of tipping points:

• **Polar Ice:** As polar ice melts, the remaining ice becomes less structurally sound and less sunlight is reflected back into space, so warming happens more rapidly. Eventually, we will reach a series of tipping points where it will become impossible to stop loss of virtually all the ice in the West Antarctic, other parts of the Antarctic, and Greenland, which would raise ocean levels dramatically.[106] There is little evidence of an irreversible tipping point in the Arctic,[107] where the ice floats on the ocean. But in the Antarctic and Greenland, ice rests on land, and if it becomes structurally weak, it will inevitably flow into the ocean and be lost. As we have seen we are already near a tipping point where West Antarctica will inevitably lose most of its ice.

- **Monsoons:** The summer monsoons in India and nearby countries depend on air circulation patterns that are sensitive to warming. At some point, these circulation patterns could shut down completely, eliminating the major source of water for hundreds of millions of people.[108]

- **Amazon Rain Forest:** Warming could change the Amazon rain forest abruptly to a grassy savannah[109] because there will be less rain and a longer dry season as temperature increases.[110] The Amazon is also in danger of reaching a tipping point because of deforestation: the trees there pull water out of the ground and release it into the atmosphere, forming clouds that bring rain to the region, so there is less rain as trees are removed. About 17% of this rainforests' trees have already been removed to open up land for agriculture and grazing; if 25% are removed, there would no longer be enough rain to support the rain forest, so it would change to savannah.[111] This would release 90 billion tons of carbon dioxide into the atmosphere as the trees die and decay.[112]

- **Boreal Forests:** These are the vast northern forests that extend through Canada, Scandinavia, and Russia. Warming makes them vulnerable to forest fires and to insect infestation and other diseases that cannot tolerate lower temperatures.[113] As a result, warming could change them from forest to a type of savannah, with groves of trees in open grassland.[114] This could release 110 billion tons of carbon dioxide into the atmosphere,[115] more than the Amazon rainforest.

- **Permafrost:** Ecosystems on lands in the far north change as permafrost thaws, allowing shrubs to spread northward into the tundra, which also increases warming by reducing the amount of sunlight the permafrost reflects. At some point, this change could make it inevitable that all the permafrost will melt,[116] causing catastrophic warming, as the methane stored in the permafrost is released. Permafrost contains more carbon than the total that is in the atmosphere now, and it would release this carbon as methane, which causes much more warming than carbon dioxide.

- **Other Ecosystems:** Many other ecosystems could reach tipping points similar to those of the Amazon and boreal forests. For example, we have seen that this sort of tipping point is very close for coral reefs and the ecosystems they support. Without a sharp reduction of global warming, many terrestrial, marine and fresh-water ecosystems will collapse in the 21st century[117]—and, of course, it will be impossible to restore them because many of their species will be extinct.

Some of these changes would happen quickly after the tipping point is reached. Others could take centuries to happen. But once we reach the tipping point, the change becomes inevitable and will persist even if we bring greenhouse gas concentrations down to the preindustrial level.

We do not know with any precision where these tipping points are, but we do know that the danger of reaching them increases as warming continues.

For example, the IPCC has said that, for some level of warming between 1.5°C and 4.5°C, it would become inevitable that all the ice of Greenland would melt.[118]

Overall, the IPCC says that there is a relatively low risk of reaching most tipping points at 1.5°C to 2.0°C of warming, but substantial risk as warming increases to 3°C.[119] This means that it is probably impossible to limit warming to 3°C or 4°C: with that much temperature increase, we are very likely to reach one of the tipping points that increases warming, making it likely that we will reach more tipping points and have runaway warming.

It obviously makes sense to limit warming to avoid the risk of reaching these tipping points.

Deniers use uncertainty to argue against action, but uncertainty means that effects could be worse than expected as well as better than expected, and it means that tipping points could come sooner as well as later. Those who say that we should not try to stop climate change because of uncertainty could use the same reasoning to say that we should go ahead and play a few rounds of Russian roulette because it is not certain that we would kill ourselves.

Australia

Many of the dangerous effects of global warming have come together in Australia, giving us a preview of what is coming to the world if we do not control global warming.

As we have seen, the Great Barrier Reef, the world's largest coral reef system occupying an area of 133,000 square miles off the northeast coast of Australia,[120] has suffered from massive coral bleaching because of global warming.

In 2003-2012 Australia had its worst drought on record, and beginning in 2017 it had an even worse drought[121]—the first time on record that three successive rainy seasons failed.[122] In rural Australia, some reservoirs were dry for over a year, so crops could not grow and residents depended on water that was trucked in.[123]

Then, beginning in the late Australian spring of 2019, it had its worst hot spell on record. On December 17, 2019 it broke records by reaching an average temperature of 105.6°F (40.9°C) across the nation; the very next day, it broke that record with an average of 107.4°F (41.9°C).[124] That was the average across the nation, and many places were hotter: in the state of South Australia, temperatures reached nearly 122°F (50°C), and asphalt roads started to melt.[125] And these records were set in the late spring, before the summer began.

The combination of drought and heat created ideal conditions for forest fires to spread. The 2019-2020 fire season started early. There were hundreds of wildfires, including some in areas that had always been considered too wet to burn.[126] In the late spring, Australia's largest city, Sydney, was blanketed with smoke, pushing air pollution to twelve times the level that is considered hazardous.[127] Smaller towns were surrounded by fire; residents could not escape because fire blocked the roads, so they fled to the beaches, where they waited—suffering from the extreme heat and smoke—for ships or helicopters to come and rescue them.

As January began, the Australian Navy, in the largest peacetime evacuation in Australia's history, rescued 4000 people who had fled to the beach from the town of Malacoota, south of Sydney in the state of Victoria.[128]

Then in mid-January, in the middle of the dry season when there is not normally rain, a freak thunderstorm put out some of the fires, and also brought hailstones the size of baseballs that destroyed cars.[129] And in early February, there was heavy rain that caused flooding.[130] These storms are evidence that global warming causes many types of extreme weather, heavy rain as well as drought and heat waves, but with their help, the fires were contained.

If there had not been freak storms in the middle of the dry season, damage would have been much worse. Even with this lucky chance, the fire burned an estimated 72,000 square miles (186,000 square kilometers) of land, over 20% of Australia's forests, killing an estimated 1 billion animals.[131] Over the course of ten days, the smoke plume from these fires traveled around the world and reached Australia again.[132]

Despite the obvious fact that fires spread more rapidly when there is drought and high temperatures, there was a campaign to blame the fires on arsonists and to distract from the importance of warming. An article making this claim was the most popular feature on the web site of *The Australian*, a newspaper owned by the conservative media mogul Rupert Murdoch, and bots were posting messages on social media sites making the same claim.[133]

Tragically, at a United Nations conference in December, Australia had joined a few other nations (including the United States) to block action encouraging nations to voluntarily strengthen their targets for emission reductions.[134] Australia is the world's largest coal exporter. The Prime Minister, Scott Morrison, once brought a lump of coal into parliament and said "This is coal. Don't be afraid,"[135] and he continued to defend the coal industry as the fires devastated his country, saying that any attempt to reduce coal production would be "reckless" and "job destroying."[136]

We can only hope that the devastation will make the dangers of global warming so clear to Australians that they will elect a parliament and Prime Minister who make Australia a leader in the world's efforts to control global warming—and that the entire world will realize that we must act vigorously to control global warming in order to avoid future catastrophes much worse than those that have already hit Australia.

Part 3
The Solutions Are Available

Up to now, this book has focused on the scientific consensus about global warming. Solutions are a political question rather than a scientific one, and we cannot expect the sort of consensus from politicians that we find among scientists. But most economists agree that pricing emissions, in one way or another, is the most effective and lowest-cost solution to global warming,[137] and combining pricing with the sort of multiple carbon offsets proposed in this section obviously would reduce emissions even more rapidly.

The World Must Act

On December 12, 2015, the nations of the world adopted the Paris Agreement by consensus. They agreed that:

- The world has the goal of keeping warming well below 2°C and will try to limit warming to 1.5°C.

- Each nation will make its own plans that are its best effort to reduce emissions and will strengthen these efforts in the future.

- There will be a review of global progress every 5 years, which will inform further efforts by individual nations.[138]

The voluntary plans that individual nations have adopted are not near enough to reach the goal; if every nation fulfills its current pledge, temperatures would rise by 3°C to 4°C.[139] But the regular review is meant to provide an opportunity to update these plans over time until they reach the goal.

Yet most nations have not been fulfilling even these modest pledges. World emissions overall have been rising rather than declining. After the agreement was adopted, Brazil elected a new president, Jair Bolsonaro, who is encouraging deforestation of the Amazon rain forest. The United States elected a new president, Donald Trump, who is the only major world leader so ignorant of science that he claimed during his campaign that global warming is a "hoax,"[140] who dismantled Obama era regulations to reduce emissions and who withdrew the United States from the Paris agreement.[141] Fortunately, we now have a saner president, who will undo much of the damage that Trump did, but Trump made us lose four years.

We have seen that it is probably impossible to stabilize world climate at 3°C to 4°C of warming, because at those temperatures, we are very likely to reach tipping points that cause more warming, such as loss of boreal forests and melting of permafrost. Thus, based on nations' behavior after adopting the Paris Agreement, we are still on track for disastrous warming.

A much more aggressive approach is justified economically. An IPCC study found that the present value of the economic damage that would be avoided by 2200 if we limit warming to 1.5°C rather than 2°C is at least four to five times greater than the cost.[142] This would require reducing world carbon dioxide emissions to net zero by 2050 and cutting other greenhouse gas emissions sharply. [143]

Because this book is directed at an American audience, it will focus on reducing the United States' emissions with a plan based on putting a price on emissions and allowing what we call "multiple emission offsets." This sort of plan also applies to other industrialized nations and to the industrialized sector of the developing nations. A different approach is needed for the portions of the rural economies of the developing nations that are based on small agriculture and herding, and carbon offsets in the industrial nations could help fund it. Of course, each nation must come up with their own plans that suit their own circumstances.

The approach described here would allow rapid reduction of emissions. It is still possible to limit warming to 1.5°C, if we start soon and if the political will is there. Currently, it does not seem likely that Congress will pass the needed legislation, but there is already some bipartisan support for a market-based approach to global warming. It is possible that, after President Biden restores command-and-control regulations, business interests will realize that a market-based approach is less costly, and there will be enough support to pass it.

Pricing Emissions

The most important component of any plan to control global warming involves putting a price on carbon dioxide and other greenhouse gas emissions and increasing that price over time until emissions become so expensive that they are eliminated. Government action is also important, but there are obvious benefits to centering efforts on pricing.

Benefits of Pricing Emissions

A fee for emissions that increases over time:

- **Encourages technological innovation.** Once legislation is passed, so it is clear that the price of emissions will continually increase, there will be a strong economic incentive to develop clean methods of production to substitute for current dirty methods. Competition will bring down the cost of these clean methods, so emissions that are expensive to eliminate initially will become cheaper to eliminate over time.

- **Reduces emissions at the lowest possible cost.** Business will reduce emissions when it is cheaper than paying the fee and will pay the fee when it is cheaper than reducing emissions. Businesses will deploy new technologies when their cost goes down to less than the fee. As the fee increases, businesses will continue to reduce emissions when it can be done more cheaply than paying the fee, so they will use the lowest cost methods of reducing emissions available at any time.

- **Is easy to adjust.** If emissions are not being reduced quickly enough, we can increase the price of emissions.

- **Allows multiple carbon offsets.** These offsets can reduce emissions more quickly while reducing hardships to businesses, as we will see.

In addition to pricing our own emissions, we would also need border adjustments, such as "emission tariffs" that impose an equivalent fee on imported goods to prevent nations that do not price emissions from having an unfair competitive advantage over our businesses. These tariffs would give other nations an incentive to reduce their emissions; another benefit of pricing emissions is that it can reduce other nations' emissions in addition to our own.

Though there is a place for other government action to reduce emissions, this sort of market-based plan should be central to reducing emissions. Because getting to net-zero emissions requires a massive overhaul of the economy, it is important to do it in the way that is lowest cost.

Methods of Pricing Emissions

There are two ways of putting a price on emissions:

- **Cap-and-trade**: The government puts a cap on emissions and creates a market that lets businesses bid for allocations of the limited amount of emissions that are allowed. The allocations can be given to businesses, so those that emit less can sell their extra allocations to those that emit more. Or there can be an all-auction cap-and-trade, where businesses have to buy all their allocations from the government. In either case, the cap goes down over time, so emissions continue to go down.

- **Emissions Tax or Fee**: The government taxes emissions, and the tax goes up over time so emissions continue to go down. This can be an actual tax with the government keeping the revenue, or it can be a fee-and-dividend system with all revenues refunded to the public.

These market-based plans are about equally effective, with minor differences between them.

The method that seems most feasible politically is the fee-and-dividend plan, a fee on emissions that gives the revenue back to all taxpayers equally. Because it does not raise revenue for the government, it is called a "fee" rather than a "tax," which makes it more politically acceptable.

Because wealthier people consume more than poorer people, they generally emit more and would pay more. One study found that, under this plan, most Americans would get back more than they pay, with lower income households getting bigger benefits.[144]

A bill to adopt a fee-and-dividend system has been introduced in Congress, the Energy Innovation and Carbon Dividend Act (HR 763). This bill taxes fuels based on their carbon content at the time when they are extracted or imported, beginning at $15 per ton and increasing by $10 per ton each year, making it relatively easy to collect the tax. It includes fees on carbon-intensive products that are imported and rebates on those that are exported. It gives all revenue that is not used to administer the plan back to citizens and legal residents.[145] But it applies only to carbon dioxide emissions from fuels and would eventually have to be extended to all sources of carbon dioxide and other greenhouse gases to be fully effective.

Acid Rain: A Pricing Success Story

We have a precedent for pricing emissions. We dealt successfully with acid rain by putting a price on pollution.

In 1990, President George H. W. Bush signed an amendment to the Clean Air Act that used cap-and-trade to deal with sulfur dioxide pollution from coal power plants, which was the main source of the acid rain that was killing forests in the northeast United States and in Canada. This act lowered the cap over time, with the goal of reducing emissions to 8.5 million tons annually, cutting them by about half. If a power plant emitted more than its allocation of pollution allowances for a year, it could either buy allowances from other plants or could cut its own emissions.

This plan created an incentive to develop technological innovations that would reduce emissions at a low cost. Scrubbers that remove sulfur dioxide from the plants smokestacks improved their performance and lowered costs over twenty years. New mining techniques increased the supply of low-sulfur coal, so some plants did not have to install scrubbers.

By 2007, emissions were down to 8.5 million tons, reaching the initial goal, and by 2010, sulfur dioxide emissions were down even further, to 5.1 million tons. Emissions of nitrogen oxides and mercury were also reduced dramatically.

Because of technological innovation, pollution was reduced much more cheaply than anyone had expected. In 1990, the EPA estimated that the plan would cost coal plants $6.1 billion. By 1998, technology had already improved so much that the Electric Power Research Organization, funded by the industry, estimated that it would cost only $1.7 billion.[146]

In the same way, pricing would encourage innovations that make the cost of reducing greenhouse gas emissions lower than current estimates.

Clean Electricity

Generating electricity accounts for 28% of our current emissions. Shifting to clean electricity also lets us clean up other sectors of the economy by shifting them from fossil fuels to electric power.

Solar, Wind, and Storage

Solar and wind power, have been going down in price rapidly, and they will be the biggest source of clean electricity in the foreseeable future. The graph shows the levelized cost of generating electricity, which is the present value of the cost of generating electricity over the lifetime of a power plant, including the cost of construction, fuel, maintenance, and any other costs. We can see that the cost of generating electricity using wind or solar is already cheaper than fossil fuels.

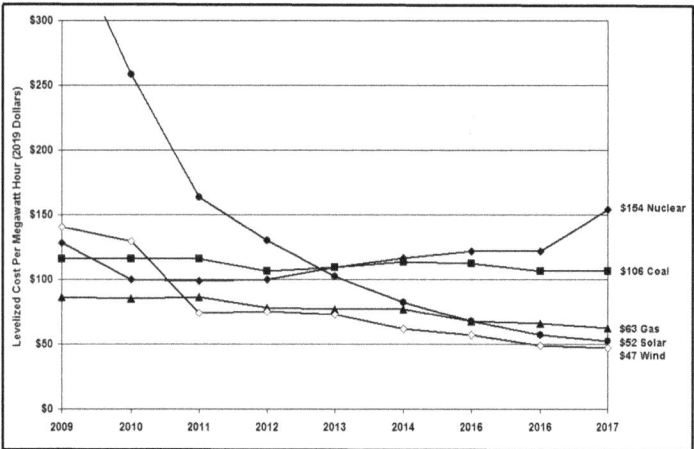

Figure 4: Levelized Cost of Generating Electricity[147]

The sun doe not always shine, and the wind does not always blow, so before we can rely on these sources for most of our electricity, we need some economical way of storing the energy so it is always available. For decades, there was speculation about what form of storage we should use, but the cost of batteries has gone down so rapidly that it is now clear that they will be the most cost-effective method of storage in the foreseeable future. Between 2012 and 2018, the cost of batteries to utilities went down 76%, from almost $800 per megawatt-hour to $187 per megawatt hour.[148] The cost of batteries is already going down more rapidly than anyone expected, and it should go down even more rapidly as we ramp up battery production to shift to electric cars.

Costs of wind and solar power differ depending on the amount of wind and sunshine where they are built. In some locations, it is already cheaper to build wind power with battery storage than any other source of electricity. Denver was stunned when it asked for bids for renewable energy with battery storage and the low bid for wind was only $21 per megawatt hour while the low bid for solar was $30 per megawatt hour. Those prices included a 30% tax credit,[149] but even without this credit, that comes come to $30 per megawatt hour for wind (less than half as much as the typical cost of natural gas) and $43 per megawatt hour for solar (less than three-quarters as much as the typical cost of natural gas). This does not include enough storage for prolonged periods with little sun and wind, which are rare, but the price of battery storage is falling so rapidly that it will ultimately be feasible to have all the storage we need.

If we put a price on emissions, making it more expensive to operate existing natural gas and coal plants, there would be a rapid shift to wind and solar with battery storage.

There would also be an incentive to invest in energy conservation and to develop technologies that bring down the cost of energy conservation to make it even cheaper than using electricity.

Government has a role in easing the transition by offering training to people who lose jobs in fossil fuel industries, preparing them for new jobs—including the many new jobs in clean energy.

Nuclear Power

We can see in the graph that the cost of nuclear power has varied from $99 to $154 per megawatt hour, much more than natural gas. The costs of wind and solar are going down, but the cost of nuclear power has not gone down since the 1970s. Because of cost, its share of global electricity generation has been declining since 1993.[150]

Even after a nuclear power plant is already built, costs of operating it are so high that one-quarter of the nuclear power plants in the United States continue to operate only because they are subsidized in order to reduce greenhouse emissions. Without the subsidy, the utilities would shift to natural gas.

For now, it makes sense to subsidize nuclear power plants that have already been built and to keep them operating (except when they pose special risks, such as being on earthquake faults or near large cities), so we can use new wind and solar plants to replace fossil fuel plants rather than nuclear plants.

But it does not make sense to build new nuclear power plants when they are so uneconomical that they are not close to being able to compete with wind and solar.

There is some talk of a new generation of nuclear reactors that will be smaller than existing reactors and will be produced in factories rather than on-site, making them cheaper than existing nuclear reactors. If this technology ever becomes commercially viable, then we will have to weigh the benefits of nuclear power against its costs: it is a source of energy that does not cause greenhouse gas emissions, but storing its radioactive wastes safely is difficult or impossible, terrorists could sabotage the plants or the fuel and waste shipments, terrorists could hijack the fuel or wastes and create dirty bombs that spread radioactivity, and there could be accidents like the ones at Fukushima or Chernobyl.

Until this new technology is available, though, there is no need to argue about nuclear power. For now, it has failed on economic grounds alone.

Clean Transportation

In the United States, transportation is the sector that causes most greenhouse gas emissions, 29% of the total. It is relatively easy to reduce the emissions of ground transportation but more difficult for air and ship transportation.

Ground Transportation

To get to net-zero emissions, we have to electrify ground transportation—cars, trucks, buses, and so on—at the same time that we shift to clean sources of electricity. We are already beginning to shift to electric vehicles, and with a price on emissions, the shift will happen more rapidly.

Electric cars are better than conventional gasoline-powered cars in many ways:

- They cost less than half as much to power,[151] largely because they use energy much more efficiently than internal combustion engines, which lose most of the gasoline's energy as waste heat.

- They are more reliable and cheaper to maintain because they have fewer moving parts.

- They accelerate better than conventional cars.

In the past, few consumers have bought them because of their high initial cost and limited range, both caused by the high cost and bulkiness of batteries. But batteries have been improving: they take less time to charge, they can store more energy in smaller batteries, and as we have seen, their cost has been going down rapidly.

Because the cost of batteries is going down faster than anyone expected, projections of future costs of electric cars have also been revised: in 2017, BloombergNEF projected electric cars would cost less than gasoline cars after 2026; but in 2018, it projected that they would cost less after 2024.[152] These are projections, and it is not possible to predict future technological developments precisely, but it does seem inevitable that electric cars will be cheaper than gasoline powered cars some time in the 2020s.

Large numbers of consumers will shift to electric cars when they cost less to buy than conventional cars. Pricing emissions would encourage people to shift to electric cars even more quickly.

Senator Chuck Schumer, Democratic leader in the Senate, has proposed a law that would make all the vehicles in America clean by 2040, which is supported by organized labor and automakers as well as environmentalists. It would give a large discount to anyone who trades in a gasoline-powered car for an electric car, with a larger discount for lower-income Americans, and it would provide grants to states to build charging stations.[153]

This law would be a perfect compliment to pricing, because it does two things that pricing cannot do:

- It helps low income people to buy electric cars, and they are least able to afford new cars.

- It builds charging stations, which is a start-up problem that it is hard for the market to solve. As long as there are not enough charging stations, most people won't buy electric cars, and as long as people don't buy electric cars, businesses are less likely to set up charging stations. Government charging stations would break this stalemate.

With both emissions pricing and this law, we could complete the shift to electric cars sooner than Schumer's goal of 2040. In addition to reducing emissions, this shift would save consumers money on fuel and would make our cities quieter and the air cleaner.

Government is also responsible for building transportation infrastructure and for zoning, and it should accommodate other low emissions forms of transportation by building transit-only lanes and protected bike lanes and by zoning to create walkable neighborhoods around transit stops.

Airplanes and Ships

We can electrify cars because drivers can stop occasionally to recharge their batteries. Airplanes and ships make much longer trips without stopping, so they need liquid fuel. It is possible to produce biofuels that are carbon-neutral but this technology is much less advanced than electric cars.

Planes and ships often travel internationally, so their emissions are not always counted in national totals though they are significant:

- **Airplanes** account for about 2.5% of global carbon dioxide emissions. With business as usual, emissions are expected to rise by at least 200% by 2050.[154]

- **Ships** account for about 3% of global emissions. With business as usual, emissions are expected to rise by 50% to 250% by 2050.[155]

Small hybrid-electric planes have been produced experimentally. SAS and Airbus are developing hybrid-electric planes for commercial use, and there are predictions that they will be commercially viable by 2032.[156] Hybrid-electric ships are already being used on very short runs, for example as ferries.[157] These technologies could reduce emissions and also make cities more livable: they would reduce noise as planes use electric propulsion during take-off and landing and reduce air pollution when ships use it near seaports.

Biofuels could get these modes down to zero emissions. They remove carbon dioxide from the air when they grow, and the fuels emit the same carbon as carbon dioxide when they are burned, so they have net-zero emissions. It is best to make biofuels from algae or seaweed, because both have high levels of oil that can be converted to fuel, and both can grow in salt water so they would not compete for farmland or fresh water.

There has been work on developing jet fuel from the Isochrysis algae, but progress is slow: biofuels accounted for only 0.01% of aviation fuel in 2018,[158] and it is estimated that biofuel from algae can only provide 3% to 5% of jet fuel by 2050.[159] Biofuel is not used for shipping, which currently uses very cheap fuel, the lowest grade produced by the oil industry, which emits so much particulate matter that it is estimated to cause 60,000 death per year; some ports protect their residents' health by requiring ships to use higher grade fuel when they are near land.[160] However, there is a company developing a ship engine powered by ammonia, which could be zero-emissions if the ammonia is produced using clean energy.[161]

If we priced emissions, there would be a faster shift to hybrids, and there would be more research and development of clean fuels. But there is no telling how long it would take these industries to shift to new technologies that have net-zero emissions.

It is a truism among economists that we cannot predict how long it will take to develop new technologies and to get their cost down to the point where they are commercially viable. As we will see, this is why we must use offsets to balance emissions that cannot be reduced quickly.

Getting to Net Zero

We have looked at the two sectors of the American economy that cause over half our greenhouse gas emissions: electric power (28% of total) and transportation (29%). We will not look in detail at industry (22%), residential and commercial (12%), or agriculture (9%), since it would be an endless task to look at all the technologies used in these sectors. But we have already seen enough to make this general observation:

- Electric cars, wind and solar power, and battery storage are already commercially viable. We could shift electrical generation and ground transportation to clean alternatives relatively quickly by putting a price on emissions to provide the economic incentive to use clean technologies and to develop innovations that drive down the prices of clean technologies even further.

- Biofuels or ammonia engines require more work before they become commercially viable and let us shift to zero-emission alternatives for air travel and shipping.

The same is true of the economy generally: clean alternatives for many sectors are available, but clean alternatives for others need more work before they become viable.

There is no way to predict how soon these new technologies will become commercially viable. If we price emissions, there will be an incentive to develop clean technologies to replace existing dirty technologies, but we cannot predict how long it will take to come up with the breakthroughs that will make it economically practical to use each of these clean technologies.

Economic Dislocation

If just a few key innovations take too long to develop, rigid emissions pricing could cause hardship and economic dislocation when prices reach high levels, possibly discrediting the entire pricing system. Here are two examples:

- **Nitrogen Fertilizer** and other synthetic fertilizers increase yield per acre; organic farming produces about 20% less per acre than conventional farming.[162] It would be good if we could develop new methods of organic farming that increase yield, but for now the fact is that, if we did not use fertilizer, we would have to convert vast areas of forest to farmland to feed the world's population, creating massive carbon dioxide emissions. For now, nitrogen fertilizer is indispensable, but it is also the main reason that nitrous oxide in the atmosphere has increased 20% since preindustrial times, making it the third most important greenhouse gas after carbon dioxide and methane.[163] There are easy ways to reduce use of nitrogen fertilizer, such as applying fertilizer when plants need it the most rather than applying massive amounts at planting time,[164] but there is no new technology on the horizon that could eliminate use of nitrogen fertilizer completely. If we put a price on emissions that increased over time, the cost of using nitrogen fertilizer could rise enough to drive up food prices and increase world hunger.

- **Cement**, which is used in concrete and mortar, is the source of about 8% of the world's carbon dioxide emissions. Cement is made by baking material that contains calcium carbonate to break it down into the calcium oxide used in cement plus carbon dioxide. We could reduce emissions by using clean energy for mining and baking, but more than half of the carbon dioxide is an unavoidable chemical by-product of breaking down calcium carbonate. There are efforts to develop clean concrete—for example, using microbes to create bio-concrete[165]—but there is no way to predict how long it will take for clean concrete to become commercially viable. For now, concrete made with cement is essential for building. If we put a price on emissions that increased over time, the high cost of cement could drive up housing prices enough to create a shortage of affordable housing.

Fertilizer and concrete are two obvious cases where a high price on emissions could create hardships, particularly for the poor, if it takes too long to develop clean technologies to eliminate emissions. There are other industrial processes, such as refining iron ore, where high prices on emissions could cause economic dislocation and unemployment if it takes too long to develop clean alternatives.

Ultimately, there will probably be innovations that will provide clean substitutes for all dirty technologies that we now use, but there is no telling how long it will take—which is one reason why any plan to price emissions should allow for emission offsets.

The IPCC has made a similar point, saying that offsets are needed to get to net-zero emissions because some non-carbon dioxide emissions "are difficult to mitigate, such as N_2O [nitrous oxide] emissions from fertilizer use and CH_4 [methane] emissions from livestock … [which] will not be reduced to zero, even under stringent mitigation scenarios."[166]

Emission Offsets

Emission offsets let businesses offset their own greenhouse gas emissions by reducing emissions in some other way. They are usually called carbon offsets, but we have seen that they are also needed to deal with emissions that contain no carbon, such as nitrous oxide. There are obvious advantages to emission offsets:

- They let us reach net zero emissions more quickly, rather than waiting until we have developed all the new clean technologies that are needed to replace the dirty technologies we use now.

- They let us reach net zero more cheaply, rather than adopting technologies whose cost is still high.

- They provide a source of funding for reducing emissions in the developing nations. For example, businesses could offset their own emissions by subsidizing farming methods in the developing nations that produce lower emissions.

- They provide an economic incentive to bring down the cost of methods that remove carbon dioxide from the atmosphere, which will be needed (as we will see) to move beyond net-zero emissions to net-negative emissions.

Multiple Emission Offsets

Multiple emission offsets can reduce emissions more quickly by letting businesses get out of paying the price for one ton of their own emissions if they reduce emissions somewhere in the world by, say, two tons or five tons.

Multiple emission offsets are economically feasible, since the costs of offsets are well below the costs that are charged by plans that put a price on emissions. For example:

- Under California's cap-and-trade plan, it costs businesses $17.45 to emit one ton of carbon dioxide equivalent.[167]

- Under the European Union's emission trading system, it costs over $25 to emit one ton of carbon dioxide from larger factories and power plants.[168]

- A recent German law begins by charging 10 Euros (about $11) to emit one ton of carbon dioxide from transportation and heating, with the price increasing to 35 Euros by 2025.[169]

By contrast, the average price in the private market for carbon offsets is $3.30 per ton,[170] though prices vary.

A government program to let business avoid paying fees would have to set standards for projects that qualify as offsets[171] and presumably would have stricter standards than some of these private programs, so let's assume that offsets would initially cost $5 per ton. And let's assume that, to give businesses an incentive to use offsets, they should cost about 80% to 90% as much as paying the fee for emissions. California could require businesses to buy offsets that reduce emissions by 3 tons to avoid paying for 1 ton of emissions, the EU could require 4.5 tons of offsets, and Germany could begin with 2 tons and work its way up to 6 tons of offsets to avoid paying the fee for 1 ton of carbon dioxide emissions.

This sort of program could jump-start global emission reductions, as businesses rush to pay for the cheapest emission reductions all over the world in order to avoid paying fees. They would still have a strong incentive to reduce their own emissions, since the offsets would cost almost as much as the fee, but they would also be reducing global emissions dramatically by paying for offsets.

It is not possible to offset all emissions. If it were, we could shift to net-zero emissions immediately, but there are obviously not enough offsets available to balance all of the world's emissions. Initially, we might let businesses offset, say, 10% of their emissions by reducing emissions anywhere in the world.

Over time, allowing offsets would have two opposite economic effects:

- Cheaper opportunities to offset emissions would be used up, driving up the price of offsets. For example, one cheap source of offsets involves sealing landfills and burning the methane that escapes, so the landfill emits carbon dioxide rather than methane. But there are only so many landfills in the world, and if a major economy let business use offsets, it would not be long before methane emissions were eliminated from all these landfills.

- Businesses would invest in developing emission-negative technologies, potentially driving down the price of offsets. Currently there is no economic incentive to develop these technologies, but once the emission reduction can be sold to businesses to use as offsets, we would expect many start-ups to begin developing these technologies and reducing their costs.

Because these two effects are opposite and because we cannot predict what technologies will be developed, we do not know whether offsets would become more or less available over time and whether their cost go up or go down.

As years go by and the price of offsets changes, governments would have to vary the multiple so businesses always have to buy offsets that cost 80% or 90% of the fees they are avoiding. As the availability of offsets changes, governments would also have to change the percent of their emissions that businesses can offset.

In the longer run, as the fee for emissions goes way up and emissions go way down, we will have to set the multiple at a level that avoids hardship and severe economic dislocation. For example, we would not want to set the price for emissions from nitrogen fertilizer so high that we drive up the price of food to the point where we cause hunger, so we would set the multiple for these emissions so offsets cost less than the full 80% or 90% of the emissions fee.

Global Cooling

As we have seen, global warming has already set off natural processes that can cause runaway warming. For example, warming has begun to melt some of the world's permafrost, which releases methane, which causes more warming, which releases more methane, and so on. Even if human emissions were net zero, global warming would continue because of these natural processes that were started by human-caused warming.

If we want world's climate to be stable, we must cool the Earth enough to stop these natural processes—which means we must go beyond net-zero emissions to net-negative emissions. This cooling can also reduce effects of global warming that we have already experienced, such as more destructive tropical storms and larger wildfires.

James Hansen, formerly director of the NASA Goddard Institute for Space Studies, has found that we must reduce carbon dioxide concentrations to 350 ppm to stabilize climate in the long term,[172] and this idea inspired the name of the environmental group 350.org. Of course, more research is needed before scientists can reach consensus about the exact level of greenhouse gases that would stabilize the climate, but it clearly must be lower than it is now.

Carbon Negative Technologies

There are a number of ways of reducing carbon dioxide concentrations that are already available or on the horizon, which would be used widely if we made them profitable by letting businesses purchase them as emissions offsets. Ultimately, after emissions are reduced so much that few offsets are needed, we will have to find other ways to fund them in order to move to net-negative emissions.

- **Trees:** 30% of the world's forest land has been cleared completely, and 20% has been degraded. Worldwide, 4.8 billion acres of forests can be restored, three-quarters of it as a mosaic that intersperses forests and agricultural uses. Planting trees can also be combined with other uses. Some crops can grow under a tree cover; shade grown coffee is the best-known example. Animals can graze on land with some trees; this is called silvapasture.[173] Currently, forest land is cleared completely for most coffee growing and grazing, but people would be eager to plant trees on this land if those trees earned them money from carbon offsets. Planting trees can also make cities more livable.

- **Restored Ecosystems:** In addition to restoring forests by planting trees, we could restore native grasslands, peatlands, kelp forests, and salt marshes, which all sequester carbon.

- **Biochar:** Biochar is produced by slowly heating agricultural wastes in the absence of oxygen, breaking them into gas and oil that can be used as fuel and into a solid residue called biochar that is rich in carbon and can be used to improve most soils and increase agricultural yield by 15% on the average, with the greatest increase in soils that are degraded or acidic.[174] Carbon dioxide is removed from the atmosphere when the plants grow, and some is sequestered in the soil rather than being released back in the atmosphere. Biochar can be produced on a large industrial scale or on a small scale at individual farms—and farmers would do this if it could earn them money from emission offsets.

- **Restored Topsoil:** Farms in America's Midwest have lost about half of the topsoil they had originally. Worldwide, about one-third of agricultural land is degraded,[175] and about 1 billion acres of farmland are so degraded that they have been abandoned.[176] The world's topsoil, composed of decayed plant material, sequesters more than twice as much carbon as the world's vegetation, and when the topsoil is lost, that carbon is released into the atmosphere as carbon dioxide.[177] If we put a price on these emissions, we would reduce farming methods that destroy topsoil and replace them with methods like no-till agriculture and conservation tillage. Farms could also sell emission offsets if they build up soil by adding compost, including biochar where appropriate. Cover crops can both prevent erosion of existing topsoil and build more topsoil by adding organic matter to the soil. This would probably be too expensive to fund exclusively with emission offsets, but it is necessary in order to conserve and restore agricultural land to feed the world's growing population without clearing more forests.

- **Biofuels Plus Carbon Capture and Storage (CCS):** CCS has been talked about mostly as a way to make fossil fuels more climate-friendly by capturing the carbon dioxide they emit and injecting it into the ground, into geological formations where it will be stored indefinitely, but the cost has not come down enough to make it competitive with wind and solar power. CCS could be carbon-negative if we generate the power using biofuels derived from algae or seaweed and store the carbon dioxide emissions underground: growing the biofuel would pull carbon dioxide out of the atmosphere, and after it is burned, the carbon dioxide would not be released into the atmosphere. We have seen that biofuels are not a mature technology yet, but their cost will come down eventually if we put a price on emissions, and the combination of the revenues from power generation and from emission offsets could make them economically viable.

- **Direct Air Capture:** Chemicals combine with carbon dioxide to capture it directly from the air; then the chemicals are stripped of the carbon dioxide so they can be reused, and the carbon dioxide is stored underground. This is obviously harder than capturing carbon dioxide from power plant emissions and is still very expensive. But the cost has fallen below $100 per ton, and the technology is attracting investors who hope the cost can be reduced enough to make it feasible.[178]

- **Enhanced Weathering of Minerals:** Silicate minerals weather naturally, combining with carbon dioxide from the atmosphere to form bicarbonates, which are ultimately washed into the ocean where the carbon is stored for hundreds of thousands of years. Enhanced weathering speeds up this process by breaking up the minerals into small pieces, so more surface area is exposed to the atmosphere and the weather. This ground silicate could be applied to farmland instead of lime to make the soil less acid. Or it could be applied to beaches, where it will wash into the ocean after weathering and counteract the acidification of the ocean caused by carbon dioxide emissions. Currently, this is expensive—$52 to $480 per ton of carbon dioxide removed from the atmosphere—but cost could come down if offsets created an incentive to bring down the price. There is currently a pilot program planned to determine the least expensive method of enhanced weathering, including the cost of mining, transporting and processing the minerals.[179]

A price on emissions with offsets would provide an incentive to bring down the cost of these methods of sequestering carbon and to develop new technologies that offset emissions.

One example is cool paving material. The United States alone has more than 2,700,000 miles of paved roads, and 94% of them are paved with asphalt mixed with aggregate particles,[180] which is black and absorbs 80% to 95% of sunlight that hits it. There has been some work on adding coatings to asphalt pavement that reflect heat into space instead of absorbing it.[181] In the long run, shifting away from fossil fuels will mean finding a substitute for asphalt, which is a petroleum product. Emission offsets would provide an incentive to develop paving materials that reflect more sunlight—though, of course, they could not be so reflective that the light impairs drivers' vision.

To give another example, there has been work on constructing six to twelve-story buildings with framing made of cross-laminated wood, which sequesters carbon assuming the wood is harvested from sustainably managed forests.[182] Wood for economic uses like this can be grown on deforested land that is so degraded that it cannot be restored to natural forests but can sequester carbon as a plantation.[183]

A price on emissions with offsets would create an incentive for developing many new technologies like these to reduce warming.

Geoengineering

But we need to look very closely at new technologies that reduce warming to see whether they have unintended consequences.

For example, there have been proposals to control warming by releasing large amounts of sulfate aerosols, which (as we have seen) reflect sunlight into space. But more detailed climate models show that, if it were done on a large enough scale and long enough to affect warming, this proposal would make some regions hotter as well as making some cooler, would cause worse drought in some regions, and would probably worsen acid rain and ozone depletion.[184] And, of course, it would not deal with acidification of the oceans or with increased carbon dioxide levels in the atmosphere. As greenhouse gas emissions kept increasing, we would have to keep releasing more and more sulfate aerosols into the atmosphere.

This sort of massive intervention in the Earth's climate is called geoengineering. Because we do not understand the world's climate completely, geoengineering could cause very destructive unintended consequences and should be avoided.

A Note on the Developing Nations

We have looked at putting a price on emissions, which is appropriate in the developed nations and in some sectors of the developing nations' economies. But it is not appropriate for the entire economies of the developing nations: many sectors cannot afford emissions fees and instead need technical advice and subsidies to reduce emissions.

We have already seen that emissions offsets could finance tree planting and building up topsoil in the developing nations. Here, very briefly, are a few other measures to reduce warming in the developing nations, which could be funded at least in part by emission offsets:

● **Clean Cookstoves:** Three billion people cook by burning wood, dung, crop residues, charcoal or coal in crude stoves or open fires, emitting carbon dioxide and black soot. Sealed cookstoves that burn these same fuels in a controlled way and send incompletely burned fuel and soot back into the stove can cut emissions by up to 95%. These stoves can also reduce the work of gathering wood and can improve health by reducing indoor air pollution from cooking, which causes over 4 million premature deaths per year.[185] Previous efforts to distribute clean cookstoves have had limited success, because the stoves did not reduce emissions sufficiently or were too hard to use for cooking.[186] New designs should be tested rigorously before they are certified to be used as emissions offsets.

● **Rice Cultivation:** Currently, rice is usually grown in flooded paddies, where organic matter decays under water without oxygen available, causing 9% to 19% of global methane emissions. We can reduce these emissions without reducing yield by draining the paddies in the middle of the season or by alternately wetting and drying the paddies, cheap methods that could be funded by emissions offsets. More radically, the System of Rice Intensification eliminates methane emissions entirely by eliminating the paddies, Instead, it plants fewer seedlings in a square grid that gives them more room to grow, waters them intermittently, and uses a rotating hoe to control weeds and aerate the soil; 4 to 5 million farmers now use this method and produce

50% to 100% more rice per acre than farmers using conventional methods.[187]

- **Grazing:** The Serengeti plain of Africa has tall grass and deep topsoil that sequester vast amount of carbon, despite all the wildlife grazing there. By contrast, commercial grazing damages the grass and depletes the topsoil by letting cattle graze the same land indefinitely. There is a movement to manage commercial grazing so it is more like wildlife grazing. The herds are moved from one area to another and kept in each area using fences, turning over the soil with their hooves and fertilizing it with their wastes while they are there; then they are kept away from that area long enough for the grass to recover. As a result, tall native grasses reestablish themselves and the topsoil gradually is restored. American farmers who use this method say that the grass is so much healthier that they can graze 200% to 300% more cattle on the same land.[188]

These practices plus incentives that prevent deforestation and loss of topsoil could dramatically reduce emissions from the rural economies of the developing nations. If emissions offsets also funded efforts to plant trees and build topsoil there, these rural economies could become carbon negative.

Extremists

There are some extreme environmental groups who oppose putting a price on carbon, saying that we should ban pollution rather than letting businesses pay to pollute, or claiming that the climate emergency is too severe for us to use pricing.

They obviously are not considering that, whatever method we use, it will take time to get to a zero-emissions economy. We cannot get rid of all our cars and shift to electric cars overnight; it will take time to retool the factories and build enough electric cars. We cannot close all fossil-fuel power plants and shift to solar and wind power overnight; it will take time to build enough clean power plants to replace the dirty plants we use now.

During this transitional period, businesses will continue to pollute. We can either make them pay to pollute, or we can let them pollute for free.

We have seen the benefits of making them pay to pollute. It lets us make the transition to a clean economy at the lowest possible cost. It encourages the development of new clean technology that will lower the cost even further. With multiple offsets, it can reduce net emissions more quickly than other methods.

There are even more environmentalists who oppose emission offsets for the same reason, saying that businesses should stop polluting rather than paying to pollute.

They obviously have not considered the benefits of multiple emission offsets, which reduce net emissions much more quickly than we could without offsets. They have not considered that offsets can also be used to help fund needed changes in the developing nations that do not have other sources of funding.

And they have not considered that offsets will make it quicker and easier to move beyond net zero to a net-negative economy, because they provide an incentive to develop technologies that sequester carbon.

Though government action can often be helpful, putting a price on emissions and allowing multiple offsets should be central to efforts to control global warming. It is better economically because it reduces emissions at a lower cost. More important, it is better environmentally.

It makes no sense to say that the climate crisis is too severe for us to using pricing. If we set prices high enough and raise them rapidly enough, pricing with multiple offsets can reduce emissions more rapidly than any other method.

Part 4
The Deniers Are Ignorant

There is general scientific consensus that global warming is caused by human activity and is dangerous, reflected in the publications of the IPCC, which are based on consensus. Though a few scientists are contrarians, the consensus is widespread. Yet the public believes that there is much more debate about the science of global warming than there really is, because there are many conservative think tanks and politicians in the United States who deny established facts on ideological grounds rather than scientific grounds. This section looks at a few common talking points of these global warming deniers.

Look at This Selected Bit of Data

Deniers often cite selected bits of data that confirm their bias. For example, measurements of the Earth's surface temperature seemed to level off for a few years after 1998, so deniers began using the slogan "Global warming stopped in 1998."[189]

In fact, as the graph shows, there is an obvious upward trend in surface temperature over the decades, though they fluctuate up and down a bit from year to year. Because of a large El Niño, 1998's temperature was well above the trend line, and temperature did not reach its level again for the next four years. Deniers cherry-picked the data by taking 1998 as their baseline, and they spent those four years claiming that global warming stopped, though things would have looked different if they had taken 1997 or any other earlier year as their baseline.

Temperatures increased slowly until 2012—again, if you take 1998 as the baseline—but then temperatures began increasing rapidly again, and deniers finally had to give up the claim that warming had stopped or slowed.

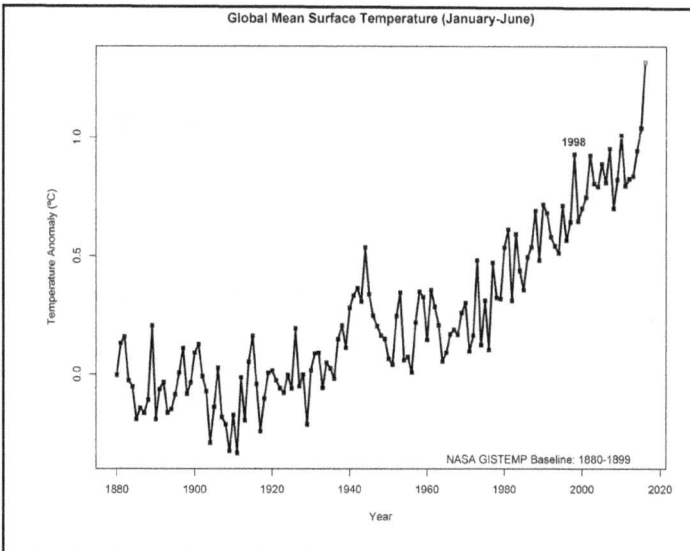

Figure 5: Average Surface Temperature[190]

Yet they keep coming up with the same sort of nonsense. Currently, they are circulating a graph of the amount of arctic ice that melted over just a few years to try to claim that melting is no longer increasing. They don't consider that it is impossible in the long run for global temperatures to keep increasing without more ice melting.

Of course, things would look very different if they showed a graph of how much ice melted since the 1970s. It would be obvious that the few years they have selected are just another temporary fluctuation.

Climate Always Changes Naturally

Deniers often claim that climate has always changed in the past because of natural processes, so the same thing must be happening today.

It's a Natural Cycle

In the 1990s, when there was still real scientific debate about the subject, some conservative scientists claimed that warming was the result of a natural cycle of variation in solar radiation, but this claim was quickly disproven. Solar radiation would warm all of the atmosphere, but in reality, warming has occurred in the lower portion of the atmosphere, the troposphere, and not in the upper portion, the stratosphere, which is exactly what we would expect as the result of warming caused by greenhouse gas emissions.[191] This idea has now been abandoned, since solar radiation has been declining while global temperatures continue to rise.

Some deniers still claim that carbon dioxide levels are changing as the result of a natural cycle, but we know this is false for two reasons:

First, the changes happening now are too fast to be explained by natural cycles. During the alternation between ice ages and warmer interglacial periods, it took thousands of years for carbon dioxide concentrations to change by 110 ppm. Now they could change by that amount in a hundred years—at the same time as human activity is emitting huge amounts of carbon dioxide. It should be very obvious that carbon dioxide is increasing so much more rapidly than it does naturally because of human-caused emissions.

Second, we can trace the origin of carbon dioxide in the atmosphere to fossil fuels based on the isotopes of carbon they contain. Carbon-12 is the most common isotope, but the atmosphere also includes small amounts of carbon-13 and carbon-14:

• Carbon-13 is less abundant in living organisms than in the atmosphere because organisms incorporate carbon-12 more readily, so it is also less abundant in fossil fuels, which are derived from living organisms. If the carbon dioxide in the atmosphere were increasing because of burning fossil fuels, we would expect the ratio of carbon-13 to be declining, which is exactly what is happening.

- Carbon-14 is radioactive. It is produced by cosmic rays striking the upper atmosphere, and it is incorporated in plants as they use the carbon dioxide in the air to grow. Because it decays over time, there is none left in fossil fuels. If the carbon dioxide in the atmosphere were increasing because of burning fossil fuels, we would expect the ratio of carbon-14 to be declining, which is exactly what is happening.[192]

For these reasons, it is certain that rising levels of carbon dioxide are not just the result of natural processes.

Climate Always Changes

The deniers' claim that climate has always changed in the past is not comforting, because these past climate changes made the Earth either more or less friendly to human life—and the changes that are occurring now are making it less friendly to human life.

Ninety million years ago, during the Cretaceous Thermal Maximum, carbon dioxide concentrations rose to over 1000 ppm, and the world was so warm that there was no polar ice and sea levels were 60 meters (200 feet) higher than they are today. This was a great climate for dinosaurs and for the plants that existed at the time, which had evolved to be adapted to heat. But much of the Earth would have been too hot for humans if humans had existed back then, and too hot to grow the crops that humans depend on for food.

The high level of carbon dioxide in the air was reduced over geologic time, as carbon dioxide was absorbed by plants, and some of these plants decayed underground, so the carbon was removed from the atmosphere and stored underground. Ultimately, reduced carbon dioxide brought a cooler climate that is more friendly to humans, to the crops we eat, and to the other plants and animals that exist today.

When we burn fossil fuels or melt the permafrost, we are releasing that stored carbon into the atmosphere, moving back toward the climate that existed in the age of dinosaurs.

Yes, climate changed in the past as deniers say. But they don't mention that, at different times in the past, the Earth's climate was more or less able to support human life—and that it is now changing in ways that make it less able to support human life.

It's Impossible to Have No Emissions

Sometimes deniers claim that it is impossible to have zero emissions, because essential industrial processes produce carbon dioxide. Sometimes they even say that we can't eliminate emissions because we emit carbon dioxide whenever we breathe.

They are showing that they do not know the meaning of the word "net" in "net-zero emissions."

Breathing involves net-zero emissions. When food is grown, it removes carbon dioxide from the air. When we breathe, we return that carbon to the air by exhaling carbon dioxide.

Even while essential industrial processes are still producing emissions, as we have seen, we can use emissions offsets to reach net-zero emissions. With a fee on emissions, we can expect that new emission-free technologies will be developed for most of these processes, but we can use offsets to get to net-zero before all of the emission-free technologies are available.

The IPCC has looked at scenarios that go beyond net-zero emissions to negative emissions—where, on the balance, human activity is removing carbon dioxide from the air. They say that it will be impossible to eliminate some emissions, such as nitrous oxide from nitrogen fertilizer and methane from livestock, but that this does not prevent us from using offsets to get to net-zero and net-negative emissions.[193]

There's Too Little Carbon Dioxide

Deniers sometimes claim that carbon dioxide from human sources is such a small portion of the atmosphere that it cannot have a large effect. In a remarkable display of ignorance, Michele Bachmann, former member of Congress and Republican presidential aspirant, said in a speech to the House of Representatives, "If carbon dioxide is a negligible gas and it's only three percent of Earth's atmosphere, what part is human activity? Human activity contributes perhaps three percent of the three percent."[194]

Other deniers, who at least know the actual amount of carbon dioxide in the atmosphere, say that a difference of 100 or 200 parts per million is such a small amount that it cannot make a big difference in the climate.

As we have seen, carbon dioxide now makes up over 410 parts per million of the atmosphere, which is less than one-twentieth of one percent; when Bachmann said it was "only 3%," she showed her ignorance of the most basic fact about climate change. And we have seen that human activity increased the concentration of carbon dioxide from 280 ppm before the industrial revolution to about 410 ppm today, so more than 30% of the carbon dioxide in the atmosphere results from human activity. Bachman showed her ignorance again when she said it was 3%.

The deniers who are a bit less ignorant than Bachmann and claim that 100 or 200 parts per million cannot make a big difference obviously do not know that a variation of just 110 ppm between the ice ages and the warm interglacial periods caused dramatic changes in climate, so 100 or 200 ppm added by human activity clearly can make a difference.

The reason is that over 99% of the atmosphere is made up of nitrogen, oxygen and argon, which do not trap heat, and about half a percent is made up of water vapor which amplifies warming but does not cause warming. Less than half a percent of the atmosphere causes warming, and a fraction of that can make a big difference.

It's Cold and Snowing Today

Whenever there is a heavy snowstorm or a cold spell, we can count on deniers to say it means that there is no global warming.

To give an extreme example, Sen. James Inhofe of Oklahoma, chair of the Senate Environment and Public Works Committee, who is known for saying that climate change is "the greatest hoax ever perpetrated on the American people," brought a snowball to the floor of the Senate on Feb. 26, 2015 and said, "we keep hearing that 2014 has been the warmest year on record." Then, taking the snowball out of a plastic bag, he said, "I ask the chair, you know what this is? It's a snowball, and that's just from outside here, so it's very, very cold out, very unseasonable." Then he threw the snowball toward the chair.[195]

People who say this sort of thing are making so many errors that it is hard to list them all.

- They confuse weather, which is the day-to-day condition of the atmosphere and is very variable, with climate, which is the weather of a place averaged over a longer period of time and is much less variable.

- They confuse local conditions with global conditions. At the same time that Inhofe was saying that it was cold in Washington DC, it was 89°F (31.7°C) in a town in Florida, very hot for February.[196] The average global temperature can be higher than usual even if your local temperature is lower than usual.

- They don't know that global warming causes heavier snowstorms. We have seen that global warming causes larger storms with more precipitation. During the winter, that precipitation is going to come down as snow. We can hope that we stop warming before temperatures go up so much that it rains in Washington DC in February rather than snowing.

- They don't know that global warming affects air currents, which can bring cold arctic air south. We have seen that there is a larger meander in the jet stream, so it can bring arctic air further south than it used to.

All of these basic facts are obviously beyond the ken of the man who chaired the Senate committee that deals with the environment.

It Would Ruin the Economy

Deniers often say that controlling global warming will devastate the economy.

American conservatives seem to believe that the market can do anything efficiently—with just one exception: it cannot produce clean energy efficiently.

During the first decade of this century, some environmentalists claimed that the world was approaching "peak oil"—the point where petroleum resources are so depleted that production inevitably declines, pushing prices up. Conservatives replied that the market would react by developing new technologies to produce more oil. And the conservatives were right: new technologies such as hydraulic fracturing ("fracking") produced so much oil that, in 2019, the United States became a net petroleum exporter for the first time since record keeping began.[197]

But conservatives won't admit that the market can do the same thing for clean energy. If we set up the right incentives by putting a price on greenhouse gas emissions that increases over the years, the market will respond with innovations that produce a cheap and abundant supply of clean energy and other clean technologies.

Of course, this change will bring some displacement, as new technologies always do. When the automobile became popular, people who raised horses and made buggies lost their jobs, and when clean energy becomes popular, people who produce dirty energy will lose their jobs. The government should respond by offering them training in new job skills and help in relocating if necessary.

But if we fail to adopt new technologies, our economy will stagnate. The United States now invests about $50 billion per year in clean energy,[198] while China invests over $125 billion in clean energy, 45% of the world total,[199] because it plans to dominate this industry.[200] If the United States does not do more to stimulate the development of clean energy, we will lose our chance to be a leader in a key industry of the twenty-first century.

It's a Hoax to Get Scientists Grants

Deniers often say that climate scientists talk about global warming so they can get grants to study it. Donald Trump and other deniers have said many times that global warming is a hoax.

How can anyone say that warming is a hoax? We can all look at pictures showing that the arctic has half as much summer ice as it did decades ago, that Glacier National Park has lost 37% of its glaciers,[201] and that an Antarctic ice shelf has collapsed. Some symptoms of global warming may be hard to understand, but it should be very obvious to everyone that all this melting ice is a sign that the Earth is getting warmer.

How can anyone say that climate scientists are perpetuating this hoax to get grants? Much of their data comes from government sources, such as NASA and NOAA. Their articles in scientific journals are peer reviewed before they are published—which means that other climate scientists read them to see if there are any weak points in the data or the analysis. To perpetuate a hoax, there would have to be a conspiracy between the climate scientists who work for the government, the climate scientists who are writing the articles, and the climate scientists who are called on to do peer review—and not a single climate scientist among them all who is honest enough to be the whistle blower who reveals this conspiracy.

And they all would be perpetuating this hoax to get grants even though any climate scientists doing work that seriously challenged the scientific consensus about global warming could get much larger grants from fossil fuel companies.

If the world gets serious about controlling global warming, the fossil fuel companies would have to leave substantial reserves in the ground, worthless stranded assets, so they are willing to spend huge amounts on global warming denial. From 2002 to 2010, the Donors Trust and Donors Capital Fund gave almost $120 million to groups that question the need to control greenhouse gas emissions; because these two groups are registered as charities, they do not have to reveal their donors—but we can guess who some of those donors are.[202] From 2013 to 2017, the Koch Family Foundation, funded by the owners of the world's largest private oil company, gave almost $55 million to the same cause.[203]

All of these donations have a limited effect because these groups cannot undermine the established science of climate change. Imagine how much they would spend to support scientists whose research showed there was some real scientific basis for denying climate change.

In an age of crazy conspiracy theories, this is one of the craziest. How can anyone possibly believe that climate change is a hoax by scientists who want to get grants and that not a single scientist has exposed this hoax, even though any scientist who could discredit the idea of climate change would get much, much larger grants from big oil?

There's Doubt So Don't Act

The deniers spread doubt to prevent us from acting against global warming, but even if their talking points had some validity, they would not justify inaction.

Imagine that you had a disease and ten doctors examined you. Nine of the doctors said you needed an operation as soon as possible to avoid being severely disabled for life, and one doctor said you didn't need the operation. Despite the doubts that the one doctor raised, anyone with common sense would not delay the operation, since nine doctors said that delay would make it too late to help.

That is something like the situation we face with global warming. The consensus of the world's climate scientists, as expressed by the IPCC, is that we must begin to reduce greenhouse gas emissions sharply as soon as possible, in order to avoid the worst effects of global warming by limiting it to 1.5°C or 2.0°C. The world's governments adopted this goal by consensus when they passed the Paris Agreement in 2015.

In the analogy, a doctor raised doubts, but in the case of climate change, those raising doubts are primarily conservative economists and politicians who are ideologically opposed to government and committed to the free market. They do not care about the science—and we have seen how little politicians like James Inhofe and Michelle Bachman know about the science.

If nine doctors said we needed the operation without delay and one said we didn't, we would have the operation. But the case of global warming is more like nine doctors saying we need the operation and one health insurance representative saying we don't because he doesn't want his insurance company to pay for it.

The science is so well established that the deniers have resorted to suppressing it rather than arguing against it. The Trump administration removed much of the information about climate change from government web sites,[204] forbade Environmental Protection Agency scientists from speaking at a conference on global warming, and disbanded independent scientific review boards.[205] The administration even rebuked weather forecasters for stating the fact that a hurricane would not affect Alabama after

Trump mistakenly claimed that it would—evidence that we have entered an age of "post-truth politics."

Our children and grandchildren will look back on us with loathing if we ignore the facts, listen to the science-denying ideologues, and delay acting on global warming until we reach tipping points that make it too late to preserve a livable world.

Notes

1: Figure 1 Source: US Global Change Research Program, https://www.globalchange.gov/browse/multimedia/global-temperature-and-carbon-dioxide

2: For more detailed information, see Rebecca Lindsey, "Climate and Earth's Energy Budget," NASA Earth Observatory, Jan 14, 2009. https://Earthobservatory.nasa.gov/features/EnergyBalance

3: IPCC, 2014: *Climate Change 2014: Synthesis Report*. Contribution of Working Groups I, II and III to the Fifth Assessment Report of the Intergovernmental Panel on Climate Change [Core Writing Team, R.K. Pachauri and L.A. Meyer (eds.)]. IPCC, Geneva, Switzerland, 2014 pp. 44-45. https://www.ipcc.ch/report/ar5/syr/

4: Figure 2 Source: NASA Earth Observatory, "Changes in the Carbon Cycle." 2011. https://Earthobservatory.nasa.gov/features/CarbonCycle/page4.php

5: Figure 3 Source: Lawrence Berkeley National Laboratory and Dept. of Energy Office of Science Biological and Environmental Research. Carbon Dioxide Information Analysis Center. https://cdiac.ess-dive.lbl.gov/

6: IPCC, 2014: *Climate Change 2014: Synthesis Report*, p. 45.

7: IPCC, 2014: *Climate Change 2014: Synthesis Report*, pp. 44-45,

8: IPCC, 2014: *Climate Change 2014: Synthesis Report*, p. 87 and pp. 44-45. United States Environmental Protection Agency, Understanding Global Warming Potentials, https://www.epa.gov/ghgemissions/understanding-global-warming-potentials.

9: Michael Mann, Climate Change: The Science and Global Impact, https://courses.edx.org/courses/course-v1:SDGAcademyX+CCSI001+3T2019/course/, Module 5.2 Stabilizing CO2 Concentrations.

10: NASA, "Atmospheric aerosols: What are they, and why are they so important?" Aug. 1, 1996, https://www.nasa.gov/centers/langley/news/factsheets/Aerosols.html Mann, Climate Change: The Science and Global Impact, Module 5.2 Stabilizing CO2 concentrations.

11: IPCC, 2014: *Climate Change 2014: Synthesis Report*, p. 84.

12: Jeff Tollefson, "Soot a major contributor to climate change: Black carbon could result in twice as much global warming as previously estimated," *Nature: International Weekly Journal of Science*, 15 January 2013. https://www.nature.com/news/soot-a-major-contributor-to-climate-change-1.12225

13: Chris Mooney, "The Arctic Ocean has lost 95 percent of its oldest ice—a startling sign of what's to come," *Washington Post*, Dec. 11, 2018. https://www.washingtonpost.com/energy-environment/2018/12/11/arctic-is-even-worse-shape-than-you-realize/

14: IPCC, 2014: *Climate Change 2014: Synthesis Report*, p. 67.

15: The International Union for Conservation of Nature (IUCN), "Peatlands and climate change," https://www.iucn.org/resources/issues-briefs/peatlands-and-climate-change

16: IPCC, 2014: *Climate Change 2014: Synthesis Report*, p. 62.

17: Kevin Schaefer, "Methane and frozen ground," National Snow and Ice Data Center. https://nsidc.org/cryosphere/frozenground/methane.html

18: Wikipedia, "John Tyndall." https://en.wikipedia.org/wiki/John_Tyndall

19: Wikipedia, "Instrumental temperature record," using NASA data. https://en.wikipedia.org/wiki/Instrumental_temperature_record#Warmest_decades

20: IPCC, 2014: *Climate Change 2014: Synthesis Report*, p. 48.

21: IPCC, 2018: Summary for Policymakers. In: *Global Warming of 1.5°C. An IPCC Special Report on the impacts of global warming of 1.5°C above pre-industrial levels and related global greenhouse gas emission pathways, in the context of strengthening the global response to the threat of climate change, sustainable development, and efforts to eradicate poverty* [Writing team: Masson-Delmotte, V., P. Zhai, H.-O. Pörtner, D. Roberts, J. Skea, P.R. Shukla, A. Pirani, W. Moufouma-Okia, C. Péan, R. Pidcock, S. Connors, J.B.R. Matthews, Y. Chen, X. Zhou, M.I. Gomis, E. Lonnoy, T. Maycock, M. Tignor, and T. Waterfield (eds.)]. p12. https://www.ipcc.ch/sr15/chapter/spm/

22: Dim Coumou, Alexander Robinson, Stefan Rahmstorf, "Global increase in record-breaking monthly-mean temperatures," *Climatic Change*, June 2013, Volume 118, Issue 3–4, pp 771–782. https://link.springer.com/article/10.1007/s10584-012-0668-1

23: Shaoni Bhattacharya, "The 2003 European heatwave caused 35,000 deaths" *New Scientist*, Oct. 10, 2003. https://www.newscientist.com/article/dn4259-the-2003-european-heatwave-caused-35000-deaths/#ixzz67jM2oM3P

24: Steve Gutterman, "Heat, smoke sent Russia deaths soaring in 2010," Reuters, October 25, 2010. https://www.reuters.com/article/us-russia-heat-deaths/heat-smoke-sent-russia-deaths-soaring-in-2010-govt-idUSTRE69O4LB20101025

25: Shekhar Chandra, "Are parts of India becoming too hot for humans?" CNN, July 3, 2019. https://www.cnn.com/2019/07/03/asia/india-heat-wave-survival-hnk-intl/index.html

26: Eun-Soon Im, Jeremy S. Pal, Elfatih A. B. Eltahir, "Deadly heat waves projected in the densely populated agricultural regions of South Asia," *Science Magazine*, Aug. 2 2017. https://advances.sciencemag.org/content/advances/3/8/e1603322.full.pdf

27: Kendra Pierre-Louis, "Heat waves in the age of climate change: Longer, more frequent and more dangerous," *New York Times*, July 18, 2019. https://www.nytimes.com/2019/07/18/climate/heatwave-climate-change.html

28: Andy Kiersz, "This heat wave is going to make you—and the rest of America—less productive, by as much as 28%," *Business Insider*, Jul 19, 2019. https://www.businessinsider.com/heat-wave-effects-on-economic-productivity-2019-7

29: IPCC, 2014: *Climate Change 2014: Synthesis Report*, p. 69.

30: IPCC, 2014: *Climate Change 2014: Synthesis Report*, p. 70.

31: IPCC, 2018: *Global Warming of 1.5°C*, p. 15-2.

32: Mann, Climate Change: The Science and Global Impact, Module 2.3, The oceans.

33: Wikipedia, "Hurricane Katrina." https://en.wikipedia.org/wiki/Hurricane_Katrina

34: Wikipedia, "Hurricane Harvey." https://en.wikipedia.org/wiki/Hurricane_Harvey

35: Wikipedia, "Hurricane Patricia." https://en.wikipedia.org/wiki/Hurricane_Patricia

36: Wikipedia, "Hurricane Irma." https://en.wikipedia.org/wiki/Hurricane_Irma

37: Wikipedia, "Cyclone Winston." https://en.wikipedia.org/wiki/Cyclone_Winston

38: Wikipedia, "Hurricane Maria." https://en.wikipedia.org/wiki/Hurricane_Maria

39: Wikipedia, "Hurricane Maria." https://en.wikipedia.org/wiki/Hurricane_Sandy

40: Wikipedia, "Hurricane Maria." https://en.wikipedia.org/wiki/Hurricane_Wilma

41: Wikipedia, "Hadley cell." https://en.wikipedia.org/wiki/Hadley_cell

42: Mann, Climate Change: The Science and Global Impact, Module 2.4 Extreme Weather and Module 7.4: Shifting Water and Food Resources.
IPCC, 2014: *Climate Change 2014: Synthesis Report*, p. 60.

43: IPCC, 2014: Climate Change 2014: Synthesis Report, p. 67.

44: Simon Michael Papalexiou Alberto Montanari "Global and regional increase of precipitation extremes under global warming," *Water Resources Research*, May 9 2019 https://doi.org/10.1029/2018WR024067

45: *The Tribune* (Chandigarh), "Torrential rain more frequent with global warming: Study," Dec 11, 2019. https://www.tribuneindia.com/news/archive/torrential-rain-more-frequent-with-global-warming-study-783506

46: Wikipedia, "East Africa Floods." https://en.wikipedia.org/wiki/2018_East_Africa_floods

47: Rael Ombuor "Inescapable effects of climate change jeopardize livelihoods across East Africa," *Voice of America News*, December 23, 2019. https://www.voanews.com/africa/inescapable-effects-climate-change-jeopardize-livelihoods-across-east-africa

48: Anna Funk, "Record rain is drowning fields in the Midwest—Is it climate change?" *Discover Magazine*, June 11, 2019. https://www.discovermagazine.com/environment/record-rain-is-drowning-fields-in-the-midwest-is-it-climate-change
Wikipedia, "2019 Midwestern U.S. floods" https://en.wikipedia.org/wiki/2019_Midwestern_U.S._floods

49: Justin Worland, "How climate change could make extreme rain even worse," *Time Magazine*, December 5, 2016. https://time.com/4591268/climate-change-rain-global-warming/. Prein, A., Rasmussen, R., Ikeda, K. et al. "The future intensification of hourly precipitation extremes," *Nature Climate Change*, Dec 5, 2106, vol. 7, pp. 48–52. https://www.nature.com/articles/nclimate3168

50: Richard A. Betts, Lorenzo Alfieri, Catherine Bradshaw, John Caesar, Luc Feyen, Pierre Friedlingstein, Laila Gohar, Aristeidis Koutroulis, Kirsty Lewis, Catherine Morfopoulos, Lamprini Papadimitriou, Katy J. Richardson, Ioannis Tsanis and Klaus Wyser, "Changes in climate extremes, fresh water availability and vulnerability to food insecurity projected at 1.5C and 2C global warming with a higher-resolution global climate model," *Philosophical Transactions of the Royal Society A*, Apr. 2, 2018. https://royalsocietypublishing.org/doi/full/10.1098/rsta.2016.0452/ Travis Aerenson, Claudia Tebaldi, Ben Sanderson and Jean-François Lamarque, "Changes in a suite of indicators of extreme temperature and precipitation under 1.5 and 2 degrees warming," *Environmental Research Letters*, Mar. 13 2018. https://iopscience.iop.org/article/10.1088/1748-9326/aaafd6

51: Henry Fountain, "Researchers Link Syrian Conflict to a Drought Made Worse by Climate Change," *New York Times*, March 2, 2015. https://www.nytimes.com/2015/03/03/science/Earth/study-links-syria-conflict-to-drought-caused-by-climate-change.html

52: Mann, Climate Change: The Science and Global Impact, Module 2.4: Extreme weather.

53: David McKenzie and Brent Swails, "'If the climate stays like this, we won't make it' say those on the frontline of Africa's drought," WPTV News, https://www.wptv.com/news/world/if-the-climate-stays-like-this-we-wont-make-it-say-those-on-the-frontline-of-africas-drought. Mike Mwenda, "The drought in Zambia is causing starvation, a power crisis and threatening the Victoria Falls," Lifegate Radio News, Dec 15, 2019. https://www.lifegate.com/people/news/drought-in-zambia-starvation-kariba-victoria-falls

54: This prediction applies to most of Africa, Australia, southern Europe, southern and central United States, northwest China, and parts of South America.
G. Naumann L. Alfieri K. Wyser L. Mentaschi R. A. Betts H. Carrao J. Spinoni J. Vogt L. Feyen, "Global changes in drought conditions under different levels of warming," *Geophysical Research Letters*, American Geophysical Union, Mar. 26, 2018. https://agupubs.onlinelibrary.wiley.com/doi/full/10.1002/2017GL076521

55: IPCC, 2018: Summary for Policymakers. In: *Global Warming of 1.5°C*. p. 9.

56: Nick Watts, Markus Amann, Prof Nigel Arnell, Sonja Ayeb-Karlsson, Kristine Belesova, Prof Maxwell Boykoff, et al. "The 2019 report of The Lancet Countdown on health and climate change: ensuring that the health of a child born today is not defined by a changing climate," *The Lancet*, Nov. 13, 2019. https://www.thelancet.com/journals/lancet/article/PIIS0140-6736(19)32596-6/fulltext
Kendra Pierre-Louis, "Climate change poses threats to children's health worldwide," *New York Times*, Nov. 13, 2019. https://www.nytimes.com/2019/11/13/climate/climate-change-child-health.html

57: Wikipedia, Fort McMurray Wildfire. https://en.wikipedia.org/wiki/2016_Fort_McMurray_wildfire

58: Robinson Meyer, "California's Wildfires Are 500 Percent Larger Due to Climate Change," *The Atlantic*, July 16 2019. https://www.theatlantic.com/science/archive/2019/07/climate-change-500-percent-increase-california-wildfires/594016/ A. Park Williams, John T. Abatzoglou, Alexander Gershunov, Janin Guzman-Morales, Daniel A. Bishop, Jennifer K. Balch, Dennis P. Lettenmaier, "Observed impacts of anthropogenic climate change on wildfire in California, Earth's future, *American Geophysical Union*, July 15, 2019. https://agupubs.onlinelibrary.wiley.com/doi/full/10.1029/2019EF001210

59: Wikipedia, "2017 California wildfires." https://en.wikipedia.org/wiki/2017_California_wildfires

60: Wikipedia, "2018 California wildfires." https://en.wikipedia.org/wiki/2018_California_wildfires

61: Dr. Joel N. Myers, "AccuWeather predicts 2018 wildfires will cost California total economic losses of $400 billion," Jul. 8, 2019. https://www.accuweather.com/en/weather-news/accuweather-predicts-2018-wildfires-will-cost-california-total-economic-losses-of-400-billion/432732

62: Wikipedia, "2019 California wildfires." https://en.wikipedia.org/wiki/2019_California_wildfires and Wikipedia, "2020 California wildfires." https://en.wikipedia.org/wiki/2020_California_wildfires

63: Another reason for the large increase in extreme events is statistical. The science of statistics tells us that the distribution of random events takes the form of a bell curve. The least common events at the far ends of the bell curve are very infrequent. But if you shift the center of the bell curve just a bit, it can have a large effect on how frequently the events at an extreme occur. Thus, as the average temperature increases a small amount, the frequency of extreme events that occur at high temperatures, such as heat waves, droughts, and flooding can increase dramatically.

64: Mann, Climate Change: The Science and Global Impact, Module 2.4: Extreme Weather/ Michael Mann, "The weather amplifier: Strange waves in the jet stream foretell a future full of heat waves and floods," *Scientific American*, March 2019, pp. 43-49. https://courses.edx.org/courses/course-v1:SDGAcademyX+CCSI001+3T2019/courseware/383bffb375fc43cb81df22655ba51f1c/6884d08dbb704693a39d1dbbba055e83/1?activate_block_id=block-v1%3ASDGAcademyX%2BCCSI001%2B3T2019%2Btype%40vertical%2Bblock%40ada62eaf7c4345bd8df80440d6113f18

65: John Ismay and Vanessa Swales, "The Arctic Plunge: From Feeling Like 92 to Freezing in a Day," *New York Times*, Nov. 12, 2019. https://www.nytimes.com/2019/11/12/us/weather-cold-temperatures.html

66: IPCC, 2014: *Climate Change 2014: Synthesis Report*, p. 42.

67: Michael Mann and Thomas Toles, *The Madhouse Effect: How Climate Change Is Threatening Our Planet, Destroying Our Politics, and Driving Us Crazy*, Columbia University Press, 2016, p. 20.

68: Wikipedia, "Northwest Passage." https://en.wikipedia.org/wiki/Northwest_Passage

69: Chris Mooney, "The Arctic Ocean has lost 95 percent of its oldest ice—a startling sign of what's to come," *Washington Post*, Dec. 11, 2018.

70: Press release by the Arctic Council's Arctic Monitoring and Assessment Programme (AMAP), "Scientists warn of rapid, unexpected Arctic shifts and urge Paris Implementation," 26 Apr 2017. https://unfccc.int/news/scientists-warn-of-rapid-unexpected-arctic-shifts-and-urge-paris-implementation

71: IPCC, 2018: *Global Warming of 1.5°C*. p. 15-2.

72: Mann, Climate Change: The Science and Global Impact, Module 6.4: Melting cryosphere. Mann and Toles, *Madhouse Effect*, p. 23.

73: University of Leeds, "Greenland ice losses rising faster than expected," *Phys Org*, Dec. 10, 2019. https://phys.org/news/2019-12-greenland-ice-losses-faster.html

74: Mann, Climate Change: The Science and Global Impact, Module 6.5.

75: Scott A Kulp and Benjamin H. Strauss, "New elevation data triple estimates of global vulnerability to sea-level rise and coastal flooding," *Nature Communications*. 10 (1): 4844, 29 October 2019. doi:10.1038/s41467-019-12808-z. PMC 6820795. PMID 31664024.

76: Climate Central, Land Projected to Be Below Tideline in 2050. https://coastal.climatecentral.org/map/9/100.8958/13.6789/?theme=sea_level_rise&map_type=coastal_dem_comparison&elevation_model=coastal_dem&forecast_year=2050&pathway=rcp45&percentile=p50&retu rn_level=return_level_0&slr_model=kopp_2014. Denise Lu and Christopher Flavelle, "Rising seas will erase more cities by 2050, new research shows," *New York Times*, Oct. 29, 2019. https://www.nytimes.com/interactive/2019/10/29/climate/coastal-cities-underwater.html

Olivia Rosane, "300 million people worldwide could suffer yearly flooding by 2050" *EcoWatch*, Oct. 30, 2019. https://www.ecowatch.com/sea-level-rise-predictions-2641159739.html

77: IPCC, 2014: *Climate Change 2014: Synthesis Report*, p. 66.

78: IPCC, 2014: *Climate Change 2014: Synthesis Report*, p. 67.

79: O. Hoegh-Guldberg, D. Jacob, M. Taylor, T. Guillén Bolaños, M. Bindi, S. Brown, et al. "The human imperative of stabilizing global climate change at 1.5°C," *Science Magazine*, American Association for the Advancement of Science, Sept.20, 2019. https://science.sciencemag.org/content/365/6459/eaaw6974

80: *Climate Interpreter*, "The Chemistry Of Ocean Acidification," Dec. 20, 2018. https://climateinterpreter.org/content/chemistry-ocean-acidification

81: IPCC, 2014: *Climate Change 2014: Synthesis Report*, p. 41.

82: James C. Orr et al., "Anthropogenic ocean acidification over the twenty-first century and its impact on calcifying organisms," *Nature*, Sept. 29, 2005. https://www.nature.com/articles/nature04095

83: Kendra Pierre-Louis, "Waters off California acidifying faster than rest of oceans, study shows," *New York Times*, Dec. 16, 2019. https://www.nytimes.com/2019/12/16/climate/california-ocean-acidifying.html

84: Wikipedia, "Ocean acidification." https://en.wikipedia.org/wiki/Ocean_acidification

85: This is the projection for RCP 2.6, which is roughly equivalent to limiting warming to 2 degrees C. IPCC, 2014: *Climate Change 2014: Synthesis Report*, p. 12.

86: Erica Goode, Climate change denialists say polar bears are fine. Scientists are pushing back," *New York Times*, April 10, 2018. https://www.nytimes.com/2018/04/10/climate/polar-bears-climate-deniers.html

87: IPCC, 2018: *Global Warming of 1.5°C*, p. 15-2.

88: IPCC, 2018: *Global Warming of 1.5°C*, p. 15-2.

89: IPCC, 2014: *Climate Change 2014: Synthesis Report*, p. 67.

90: Mann, Climate Change: The Science and Global Impact, Module 7.3 Ecosystems and biodiversity.

91: IPCC, 2018: *Global Warming of 1.5°C*, p. 15-2.
IPCC, 2018: Summary for Policymakers. In: *Global Warming of 1.5°C*, p.8.

92: IPCC, 2014: *Climate Change 2014: Synthesis Report*, p. 69.

93: Kendra Pierre-Louis, "The world is losing fish to eat as oceans warm, study finds," *New York Times*, Feb. 28, 2019. https://www.nytimes.com/2019/02/28/climate/fish-climate-change.html
Kendra Pierre-Louis, "Warming waters, moving fish: How climate change is reshaping Iceland," *New York Times*, published Nov. 29, 2019, updated Dec. 3, 2019 https://www.nytimes.com/2019/11/29/climate/climate-change-ocean-fish-iceland.html

94: Mann, Climate Change: The Science and Global Impact, Module 7.4 Shifting food resources.

95: National Academy of Science, *Himalayan Glaciers: Climate Change, Water Resources, and Water Security*, Washington DC, National Academies Press, 2012. https://www.nap.edu/read/13449/chapter/3

96: Damian Carrington, "A third of Himalayan ice cap doomed, finds report," *The Guardian*, 4 Feb 2019. https://www.theguardian.com/environment/2019/feb/04/a-third-of-himalayan-ice-cap-doomed-finds-shocking-report

97: IPCC, 2018: *Global Warming of 1.5°C*, p. 15-9.

98: IPCC, 2018: Summary for Policymakers. In: *Global Warming of 1.5°C*, p. 9.

99: Internal Displacement Monitoring Centre (IDMC) – Norwegian Refugee Council. "Displacement due to natural hazard-induced disasters: Global estimates for 2009 and 2010," June 2011. http://www.internal-displacement.org/publications/displacement-due-to-natural-hazard-induced-disasters-global-estimates-for-2009-and

100: O. Brown, "Migration and Climate Change," IOM Migration Research Series, paper no.31, 2008. https://publications.iom.int/books/mrs-ndeg31-migration-and-climate-change. N. Stern, ed., *The Economics of Climate Change: The Stern Review*, Cambridge University Press, 2006. http://mudancasclimaticas.cptec. inpe.br/~rmclima/pdfs/destaques/sternreview_report_complete.pdf. Friends of the Earth, "A Citizen's Guide to Climate Refugees, Fact Sheet Four: Predictions of Climate Refugees to 2050," Friends of the Earth, 2007. Mevlüt Çavuşoğlu, *The Problem of Environmental Refugees*, Parliamentary Assembly Doc. 11084, 23 October 2006. https://assembly.coe.int/nw/xml/XRef/Xref-DocDetails-EN. asp?FileID=19734&lang=EN

101: Henry Fountain, "Researchers link Syrian conflict to a drought made worse by climate change," *New York Times*, March 2, 2015. https://www.nytimes. com/2015/03/03/science/Earth/study-links-syria-conflict-to-drought-caused-by-climate-change.html

102: Mann, Climate Change: The Science and Global Impact, Module 7.6 Security concerns.

103: IPCC, 2014: *Climate Change 2014: Synthesis Report*, pp. 70, 73.

104: Mann, Climate Change: The Science and Global Impact, Modules 2.5 Sea ice, glaciers & global sea level, 2.6 Paleoclimate evidence, 4.6 Interpreting climate sensitivity, 6.1 Surface temperature projections, and 6.4 Melting cryosphere. Mann and Toles, *The Madhouse Effect*, pp. 59-60.

105: Martin L. Weitzman, "On modeling and interpreting the economics of catastrophic climate change," *The Review of Economics and Statistics*, Vol. XCI, Number 1, February 2009. https://scholar.harvard.edu/files/weitzman/files/ modelinginterpretingeconomics.pdf

106: Mann, Climate Change: The Science and Global Impact, Module 6.4: The Melting Cryosphere

107: IPCC, 2014: *Climate Change 2014: Synthesis Report*, p. 74.

108: Mann, Climate Change: The Science and Global Impact, Module 7.7: Tipping points.

109: IPCC, 2014: *Climate Change 2014: Synthesis Report*, p. 70.

110: Fen Montaigne, "Will deforestation and warming push the Amazon to a tipping point?" *YaleEnvironment 360*, September 4, 2019. https://e360.yale.edu/ features/will-deforestation-and-warming-push-the-amazon-to-a-tipping-point

111: Monica de Bolle, "Amazon deforestation is fast nearing tipping point when rainforest cannot sustain itself," Peterson Institute for International Economics, October 23, 2019. https://www.piie.com/research/piie-charts/amazon-deforestation-fast-nearing-tipping-point-when-rainforest-cannot-sustain

112: Timothy M. Lenton, Johan Rockström, Owen Gaffney, Stefan Rahmstorf, Katherine Richardson, Will Steffen & Hans Joachim Schellnhuber, "Climate tipping points — too risky to bet against," *Nature: A Nature Research Journal*, Nov. 27 2019. https://www.nature.com/articles/d41586-019-03595-0

113: "Boreal forest suffering 'significant' losses in Russia, Canada," CBC News, April 3, 2015. https://www.cbc.ca/news/technology/boreal-forest-suffering-significant-losses-in-russia-canada-1.3019507

114: "Boreal forest being driven to tipping point by climate change, study finds," CBC News, August 21, 2015. https://www.cbc.ca/news/technology/boreal-forest-being-driven-to-tipping-point-by-climate-change-study-finds-1.3198892?cmp=rss

115: Lenton et al, "Climate tipping points — too risky to bet against."

116: IPCC, 2014: *Climate Change 2014: Synthesis Report*, p. 70.

117: IPCC, 2014: *Climate Change 2014: Synthesis Report*, p. 74.

118: IPCC, 2014: *Climate Change 2014: Synthesis Report*, p. 72.

119: IPCC, 2018: *Global Warming of 1.5°C*, 2018, pp. 15-20.

120: Wikipedia, "Great barrier reef." https://en.wikipedia.org/wiki/Great_Barrier_Reef

121: Wikipedia, "Drought in Australia." https://en.wikipedia.org/wiki/Drought_in_Australia

122: Umair Irfan, "Australia's hellish heat wave and wildfires, explained," *Vox*, Jan. 6, 2020. https://www.vox.com/2019/12/30/21039298/40-celsius-australia-fires-2019-heatwave-climate-change

123: Livia Albeck-Ripka, "As Water Runs Low, Can Life in the Outback Go On?" *New York Times*, Dec. 8, 2019. https://www.nytimes.com/2019/12/08/world/australia/water-drought-climate.html

124: Irfan, "Australia's hellish heat wave."

125: Livia Albeck-Ripka, "2 firefighters die in Australia fires and Scott Morrison ends vacation," *New York Times*, published Dec. 19, 2019, updated Dec. 31, 2019. https://www.nytimes.com/2019/12/19/world/australia/fires-firefighters-killed-scott-morrison.html

126: Adam Morton, "Unesco expresses concern over bushfire damage to Australia's Gondwana rainforests," *The Guardian*, Nov. 28, 2019. https://www.theguardian.com/australia-news/2019/nov/28/unesco-expresses-concern-over-bushfire-damage-to-australias-gondwana-era-rainforests

127: Peter Dockrill, "Fires in Australia Just Pushed Sydney's Air Quality 12 Times Above 'Hazardous' Levels," *Science Alert*, Dec. 11, 2019. https://www.sciencealert.com/sydney-air-soars-to-12-times-hazardous-levels-under-toxic-blanket-of-bushfire-smoke

128: James Patterson, "Australia Bush Fires: Navy Beach Rescues Are Largest in Peacetime," *International Business Times*, Jan. 4, 2020. https://www.ibtimes.com/australia-bush-fires-navy-beach-rescues-are-largest-peacetime-2896198

129: Jamie Tarabay, "Australia's Wild Weather: First Fires, Now Baseball-Size

Hail," *New York Times*, Jan. 20, 2020. https://www.nytimes.com/2020/01/20/world/australia/weather-hail-sydney-canberra.html

130: "Sydney rains: Record rainfall brings flooding but puts out mega-blaze," BBC, February 10, 2020. https://www.bbc.com/news/world-australia-51439175

131: Wikipedia, "2019–20 Australian bushfire season." https://en.wikipedia.org/wiki/2019%E2%80%9320_Australian_bushfire_season
Patrick Galey, "Those Horrific Wildfires in Australia Destroyed a Fifth of The Continent's Forests, Science Alert, Feb. 25, 2020. https://www.sciencealert.com/bushfires-decimated-a-staggering-fifth-of-australia-s-unique-forests

132: Rafael Cereceda "Australia wild fire smoke goes around the world in 10 days, Euronews, Feb. 1, 2020. https://www.euronews.com/2020/01/17/australia-wild-fire-smoke-goes-around-the-world-in-10-days

133: Damien Cave, "How Rupert Murdoch is influencing Australia's bushfire debate: Critics see a concerted effort to shift blame, protect conservative leaders and divert attention from climate change," *New York Times*, Jan. 8, 2020. https://www.nytimes.com/2020/01/08/world/australia/fires-murdoch-disinformation.html

134: Somini Sengupta, "U.N. climate talks end with few commitments and a 'lost' opportunity," *New York Times*, Dec. 15, 2019. https://www.nytimes.com/2019/12/15/climate/cop25-un-climate-talks-madrid.html

135: Sonali Paul, "In coal we trust: Australian voters back PM Morrison's faith in fossil fuel," Reuters, May 19, 2019. https://www.reuters.com/article/us-australia-election-energy/in-coal-we-trust-australias-voters-back-pm-morrisons-faith-in-fossil-fuel-idUSKCN1SP06F

136: Giovanni Torre, "Scott Morrison says he will not make 'reckless' cuts to Australian coal industry amid criticism of climate change response," *The Telegraph*, Dec. 23, 2019. https://www.telegraph.co.uk/news/2019/12/23/scott-morrison-says-will-not-make-reckless-cuts-australian-coal/

137: In 2019, all four former chairs of the Federal Reserve Bank, 15 former chairs o the Council of Economic Advisors, 27 Nobel Laureate economists, and 3558 American economists signed an open letter calling for a tax on carbon emissions with revenues returned equally with all American citizens. Climate Leadership Council. https://clcouncil.org/economists-statement/

138: United Nations Climate Change, The Paris Agreement: Essential Elements. https://unfccc.int/process-and-meetings/the-paris-agreement/the-paris-agreement

139: IPCC, 2018: *Global Warming of 1.5°C*, p. 15-2.

140: Jeremy Schulman, "Every Insane Thing Donald Trump Has Said About Global Warming: Well, most of them, anyway!" *Mother Jones*, December 12, 2018. https://www.motherjones.com/environment/2016/12/trump-climate-timeline/

141: Wikipedia, "United States withdrawal from the Paris agreement." https://en.wikipedia.org/wiki/United_States_withdrawal_from_the_Paris_Agreement

142: According to the IPCC report, limiting warming to 1.5 degrees C requires getting to net-zero carbon dioxide emissions by 2059, which would require annual investment until 2050 of $1.46 to $3.51 trillion (in 2010 dollars) in energy supply and $640 to $910 billion annual investment in reducing energy demand. The present value of the damages that would be avoided by 2200 is estimated at $496 trillion. Add in other damages that are difficult to quantify, such as displacement of people and destruction of ecosystems by warming, and the present value of the benefits may be at least four to five times as great as the cost. IPCC, 2018: *Global Warming of 1.5°C*, p. 15-2.

143: IPCC, 2018: Summary for Policymakers. In: *Global Warming of 1.5°C*, p. 12.

144: Anna Cheyette, "Why A Carbon Tax Is Good For The U.S. Economy," *Wharton School Public Policy Initiative*, Nov. 1, 2016. https://publicpolicy.wharton. upenn.edu/live/news/1519-why-a-carbon-tax-is-good-for-the-us-economy

145: Summary: H.R.763 — 116th Congress (2019-2020). https://www.congress. gov/bill/116th-congress/house-bill/763

146: Justin Gerdes, "Cap and trade curbed acid rain: 7 reasons why it can do the same for climate change," *Forbes*, Feb 13, 2012.
Robert Stavins, Harvard Environmental Economics Program, "The U.S. sulphur dioxide cap and trade programme and lessons for climate policy." *Vox*, August 12, 2012. https://www.hks.harvard.edu/publications/us-sulphur-dioxide-cap-and-trade-programme-and-lessons-climate-policy

147: Calculated from Lazard's Annual Levelized Cost of Energy Analysis for 2017. https://www.lazard.com/perspective/levelized-cost-of-energy-2017/

148: H, Mai, "Electricity costs from battery storage down 76% since 2012: BNEF" *Utility Dive*, Mar. 26, 2019. https://www.utilitydive.com/news/electricity-costs-from-battery-storage-down-76-since-2012-bnef/551337/

149: Megan Geuss, "Wind with batteries? Build it quickly and it could cost $21/MWh in Colorado," *Ars Technica*, 1/19/2018, https://arstechnica.com/information-technology/2018/01/colorado-could-get-some-of-the-cheapest-wind-and-solar-in-the-us-with-caveats/

150: IPCC, 2014: *Climate Change 2014: Synthesis Report*, p. 100.

151: "Electric vehicles: Saving on fuel and vehicle costs," US Department of Energy, Office of Energy Efficiency and Renewable Energy https://www.energy. gov/eere/electricvehicles/saving-fuel-and-vehicle-costs
"Costs and benefits of electric cars vs. conventional vehicles," *Energy Sage*, 2018. https://www.energysage.com/electric-vehicles/costs-and-benefits-evs/evs-vs-fossil-fuel-vehicles.

152: Nathaniel Bullard, "Electric car price tag shrinks along with battery cost," *Bloomberg Opinion*, Apr. 12, 2019. https://www.bloomberg.com/opinion/articles/2019-04-12/electric-vehicle-battery-shrinks-and-so-does-the-total-cost
Likewise, the President of General Motors has said, "we think electric vehicle propulsion systems will achieve cost parity with internal combustion engines within a decade, probably sooner." Mark Reuss, "GM president: Electric cars won't go mainstream until we fix these problems," CNN Business Perspectives, November

25, 2019. https://www.cnn.com/2019/11/25/perspectives/gm-electric-cars/index.html

153: Chuck Schumer, "A bold plan for clean cars." *New York Times*, Oct. 24, 2019. https://www.nytimes.com/2019/10/24/opinion/chuck-schumer-electric-car.html?action=click&module=Opinion&pgtype=Homepage

154: Hiroko Tabuchi, "'Worse than anyone expected': Air travel emissions vastly outpace predictions," *New York Times*, Sept. 19, 2019. https://www.nytimes.com/2019/09/19/climate/air-travel-emissions.html

155: Paul Hawken, ed., *Drawdown: The Most Comprehensive Plan Ever Proposed to Reverse Global Warming*, Penguin Books, 2017, p. 140.

156: Marisa Garcia, "Could hybrid electric planes fly commercial? Airbus and SAS aim to find out." *Forbes*, May 29, 2019. https://www.forbes.com/sites/marisagarcia/2019/05/29/could-hybrid-electric-planes-fly-commercial/#347126f052fc.
Wikipedia, "Hybrid electric aircraft" https://en.wikipedia.org/wiki/Hybrid_electric_aircraft

157: Wikipedia, "Integrated electric propulsion." https://en.wikipedia.org/wiki/Integrated_electric_propulsion

158: International Energy Agency, "Transport biofuels." https://www.iea.org/tcep/transport/biofuels/

159: Wikipedia, "Aviation biofuel," https://en.wikipedia.org/wiki/Aviation_biofuel

160: Hawken, *Drawdown*, p. 141.

161: Roger Harrabin, "Climate change: Fertiliser could be used to power ocean-going ships," BBC, February 19, 2020. https://www.bbc.com/news/business-51548361

162: Lauren C. Ponisio, Leithen K. M'Gonigle, Kevi C. Mace, Jenny Palomino, Perry de Valpine and Claire Kremen, "Diversification practices reduce organic to conventional yield gap," *Proceedings of the Royal Society: Biological Sciences*, Jan. 22, 2015. https://royalsocietypublishing.org/doi/full/10.1098/rspb.2014.1396

163: Robert Sanders, "Fertilizer use responsible for increase in nitrous oxide in atmosphere," *UC Berkeley News*, Apr. 2, 2012. https://news.berkeley.edu/2012/04/02/fertilizer-use-responsible-for-increase-in-nitrous-oxide-in-atmosphere/
Amos Zeeberg, "Bricks Alive! Scientists Create Living Concrete," *New York Times*, Jan. 15, 2020. https://www.nytimes.com/2020/01/15/science/construction-concrete-bacteria-photosynthesis.html

164: Hawken, *Drawdown*, p. 56.

165: Lucy Rodgers, "Climate change: The massive CO2 emitter you may not know about," BBC News, 17 December 2018, https://www.bbc.com/news/science-environment-46455844

166: IPCC, 2014: *Climate Change 2014: Synthesis Report*, p. 84.

167: Melanie Curry "Latest cap-and-trade auction shows strong results" California Streetsblog, May 22, 2019. https://cal.streetsblog.org/2019/05/22/latest-cap-and-trade-auction-shows-strong-results/
Price on Carbon, "California cap and trade" https://priceoncarbon.org/california-cap-and-trade/

168: That is the price on January 15, 2019 according to Markets Insider Watchlist. https://markets.businessinsider.com/commodities/co2-european-emission-allowances For more information on the emissions that are covered and the difficulties in starting the system, see Wikipedia, European Union Emissions Trading Scheme. https://en.wikipedia.org/wiki/European_Union_Emission_Trading_Scheme

169: Melissa Eddy, "Germany passes climate-protection law to ensure 2030 goals," *New York Times*, Nov. 15, 2019 https://www.nytimes.com/2019/11/15/world/europe/germany-climate-law.html

170: Ecosystem Marketplace, A Forest Trends Initiative, *Raising Ambition: State of the Voluntary Carbon Markets 2016*, May 26, 2016, p. 6. https://www.forest-trends.org/publications/raising-ambition/

171: California has already begun to set standards for offsetting emissions by preserving rainforests. Michael Oppenheimer and Steve Schwartzman, "How California can save the Amazon," *New York Times*, Aug. 29, 2018. https://www.nytimes.com/2018/08/29/opinion/california-climate-save-amazon.html

172: James Hansen, Makiko Sato, Pushker Kharecha, David Beerling, Valerie Masson-Delmotte, Mark Pagani, Maureen Raymo, Dana L. Royer, and James C. Zachos, "Target atmospheric CO2: Where should humanity aim?" Columbia University, April 7, 2008. http://www.columbia.edu/~jeh1/2008/TargetCO2_20080407.pdf

173: Hawken, *Drawdown*, pp. 114-115, 58-59, 50-51.

174: Hawken, *Drawdown*, pp. 64-65.

175: Thin Lei Win, "Better soil could trap as much planet-warming carbon as transport produces: study," Reuters, November 14, 2017. https://www.reuters.com/article/us-climatechange-agriculture-soil/better-soil-could-trap-as-mush-planet-warming-carbon-as-transport-produces-study-idUSKBN1DE2DB

176: Hawken, *Drawdown*, p. 41.

177: David R. Montgomery and Slayde Hawkins Dappen, "Soil, our secret weapon against climate change," *Climate Solutions*, June 17, 2013. https://www.climatesolutions.org/article/soil-our-secret-weapon-against-climate-change

178: Rory Jacobson, "The case for investing in direct air capture just got clearer," Greenbiz, May 28, 2019. https://www.greenbiz.com/article/case-investing-direct-air-capture-just-got-clearer

179: David Beerling, "How 'enhanced weathering' could slow climate change and boost crop yields" *Carbon Brief*, Feb. 19, 2018. https://www.carbonbrief.org/guest-post-how-enhanced-weathering-could-slow-climate-change-and-boost-crop-yields

Peter Fimrite "Could putting pebbles on beaches help solve climate change?" *San Francisco Chronicle*, Dec. 17, 2019. https://www.sfchronicle.com/environment/article/Could-putting-pebbles-on-beaches-help-solve-14911295.php

180: National Asphalt Pavement Association, "Engineering Overview." https://www.asphaltpavement.org/index.php?option=com_content&view=article&id=14

181: Wikipedia, "Cool pavement." https://en.wikipedia.org/wiki/Cool_pavement

182: Frank Lowenstein, Brian Donahue and David Foster, "Let's Fill Our Cities With Taller, Wooden Buildings: Trees are some of our best allies in solving the climate crisis." *New York Times*, Oct. 3, 2019 https://www.nytimes.com/2019/10/03/opinion/wood-buildings-architecture-cities.html

183: Hawken, *Drawdown*, pp. 132-134.

184: Mann, Climate Change: The Science and Global Impact, Module 8.2: Geoengineering 2.
Mann and Toles, *The Madhouse Effect*, p. 122-123.

185: Hawken, Drawdown, 44-45.

186: Sara Morrison, "Undercooked: An expensive push to save lives and protect the planet falls short," *ProPublica*, July 12, 2018. https://www.propublica.org/article/cookstoves-push-to-protect-the-planet-falls-short

187: Hawken, *Drawdown*, pp. 48-49.

188: Hawken, *Drawdown*, pp. 72-73.

189: Rebecca Lindsey, "Did global warming stop in 1998?" NOAA Climate.gov, September 4, 2018. https://www.climate.gov/news-features/climate-qa/did-global-warming-stop-1998

190: Source: NASA/Goddard Institute for Space Studies, 2016 Climate Trends Continue to Break Records, July 19, 2016. https://www.nasa.gov/feature/goddard/2016/climate-trends-continue-to-break-records

191: Mann, Climate Change: The Science and Global Impact, Module 4.5 Detecting climate change.

192: John Cook, "The human fingerprint in global warming, Skeptical Science, March 29 2010 https://skepticalscience.com/human-fingerprint-in-global-warming.html

193: IPCC, 2014: *Climate Change 2014: Synthesis Report*, p. 84.

194: Brad Johnson, "Stumped by science: Michele Bachmann calls CO2 'harmless,' 'negligible,' 'necessary,' 'natural,'" ClimateProgress, April 24, 2009. https://thinkprogress.org/stumped-by-science-michele-bachmann-calls-co2-harmless-negligible-necessary-natural-ba603ff8ef05/

195: Jeffrey Kluger, "Senator throws snowball! Climate change disproven!" *Time Magazine*, Feb. 27, 2015. https://time.com/3725994/inhofe-snowball-climate/

196: Kluger, "Senator throws snowball!"

197: Ariel Cohen, "Making history: U.S. exports more petroleum than it imports in September and October" *Forbes*, Nov 26, 2019. https://www.forbes.com/sites/arielcohen/2019/11/26/making-history-us-exports-more-petroleum-than-it-imports-in-september-and-october/#19d62d075f3b

198: Statista, "Value of investments in renewable energy in the U.S. from 2004 to 2018." https://www.statista.com/statistics/186818/north-american-investment-in-sustainable-energy-since-2004/

199: Wikipedia, "Renewable energy in China." https://en.wikipedia.org/wiki/Renewable_energy_in_China

200: Dominic Dudley, "China is set to become the world's renewable energy superpower, according to new report," *Forbes*, Jan. 11, 2019. https://www.forbes.com/sites/dominicdudley/2019/01/11/china-renewable-energy-superpower/#9c446f8745a2

201: Christina Maxouris and Andy Rose "Glacier National Park is replacing signs that predicted its glaciers would be gone by 2020," CNN, January 8, 2020. https://www.cnn.com/2020/01/08/us/glaciers-national-park-2020-trnd/index.html?utm_content=2020-01-08T11%3A35%3A42

202: Wikipedia, "Donors Trust." https://en.wikipedia.org/wiki/Donors_Trust

203: Greenpeace, "Koch Industries: Secretly funding the climate denial machine." https://www.greenpeace.org/usa/global-warming/climate-deniers/koch-industries/

204: Chris Baynes, "Trump administration removes quarter of all climate change references from government websites: Researchers warn move has 'severely weakened public access' to information about environment," *The Independent*, July 25, 2019. https://www.independent.co.uk/news/world/americas/us-politics/trump-climate-change-government-websites-global-warming-a9020461.html

205: Lisa Friedman, "Bipartisan report says Trump's abuse has pushed federal science to a 'crisis,'" *New York Times*, Oct. 3, 2019. https://www.nytimes.com/2019/10/03/climate/trump-science-crisis.html